THE OPTICAL TRANSFER FUNCTION

Monographs on Applied Optics
No. 3

The Optical Transfer Function

K. R. BARNES, M.Sc.

Rank Precision Industries Ltd

WITH A PREFACE BY

W. D. WRIGHT

Professor of Applied Optics
Imperial College of Science and Technology

American Elsevier Publishing Company, Inc.
New York

Published in the United States by
American Elsevier Publishing Company, Inc.
52 Vanderbilt Avenue, New York, New York 10017

Library of Congress Catalog Card Number 70–170296

International Standard Book Number 0–444–19592–0

Printed in Great Britain

Preface

by

PROFESSOR W. D. WRIGHT

Applied Optics Section, Imperial College of Science and Technology

The optical transfer function has attracted a great deal of attention during the past 20 years and is now an established branch of optics. Yet although no one would deny the theoretical elegance with which the performance of an optical system can be expressed in terms of its OTF, not everyone is convinced that the results give a meaningful evaluation of what a lens actually does. OTF, in fact, has its enthusiasts and its detractors.

In this situation a survey by a young and unprejudiced mind can provide a very real contribution to the literature and Mr Barnes' M.Sc. project provided him with just such an opportunity. He has now expanded his original report into the present monograph dealing not only with the underlying theory but also with the methods of OTF measurement and computation.

I have a personal interest in the subject to the extent that in the early days of television, when I was concerned with the selection of suitable lens systems for television cameras, I made the suggestion that the performance of a lens might be conveniently expressed in terms of the frequency attenuation it introduced into the overall television network. That was as long ago as 1938 (*Reports on Progress in Physics,* Vol. 5, p. 203), but ideas by themselves are not worth very much unless you do something about them. Mr Barnes' monograph illustrates very clearly the wealth of ideas and the wealth of activity that now surround the subject, but he acknowledges that more still needs to be known about the eye (p. 2) and the photographic process (p. 64), before the OTF can be used to select the best lens for certain applications. This is an indication of the balanced judgment which Mr Barnes brings to bear on his treatment of the OTF.

Author's Preface

In recent years the Optical Transfer Function (OTF) has become increasingly important in the specification and testing of lenses and complete optical systems. This has been due to the fact that the OTF provides a more complete description of the performance of an optical system than do the methods used formerly, and it also provides an important link between the terminologies of optics and electrical engineering, which are being brought ever closer together in the production of electro-optical systems. The majority of such systems operate with 'white' light, and it is for this reason that the emphasis in this monograph is on the OTF over such extended wavelength ranges.

The aim is to provide an introduction to the theory and practice of OTF measurement and calculation on which the reader may build from the large amount of detailed work which has been done on the subject over the last fifteen years, and also in the last chapter to show by some examples how the OTF may be used to predict the performance of optical systems.

The mathematics employed are not demanding, although at least an acquaintance with Fourier transform theory is required. There is also no attempt made to achieve complete mathematical rigour, which I feel would tend to hide the practically important aspects of the subject in a welter of usually trivial detail, and would be out of place here. No universally accepted notation has evolved for the OTF and its associated quantities, so in this monograph a system is used which, it is hoped, has at least some transparency.

Most of the material contained in this monograph comes from the dissertation which I produced whilst studying at Imperial College in 1968, and I would like to express my thanks to members of the Applied Optics department for their useful advice and discussions at that time. The text has been revised and brought up to date where necessary and in particular new material has been added to Chapter 5. With regard to this chapter I would like to thank the management of Broadcast Division, Rank Precision Industries, who allowed me to use their computing facilities, and Mr K. B. Sadhvani for many helpful discussions on the practical use of OTF calculation.

Finally my thanks are most due to my wife who not only assisted with some of the calculations but was also a great source of encouragement and has shown much patience.

LEICESTER
May, 1971

K. R. BARNES

Contents

1

Introduction

The optical transfer function (OTF) describes the ability of an optical system to transfer the spatial distribution of light in an object to its image. A knowledge of the OTF of the imaging system and the light distribution in the object make it possible to calculate the light distribution in the image.

This important calculation was not possible with methods of optical image evaluation used before the advent of the OTF, such as the limiting resolution of the optical system or its wavefront aberration, and the contribution of the optical part to the overall image quality of an electro-optic system, such as television, was rather uncertain. It was to overcome this difficulty that Schade (1948) and Elias *et al.* (1952) produced papers applying electrical linear network theory to the optical imaging process.

However, apparently unknown to these authors, Duffieux (1946), in a series of papers from 1935 which were printed privately, seems to have been the first to introduce Fourier transform theory to optical image evaluation. Hopkins (1951, 1953), as a result of his theoretical studies of microscope condenser systems, contributed a good deal to the early establishment of a firm theoretical basis for the OTF, and this work has been built upon to provide methods of calculating the OTF from the wavefront aberration or design data of a lens. At the same time many methods of measurement of the OTF have been developed, and the results obtained confirm theoretical predictions.

The specification and testing of lenses in terms of the OTF

are rapidly becoming accepted by both users and manufacturers, who recognize their advantages. However, a few words of warning are in order here concerning the limitations of the method.

An OTF can be a very sensitive measure of performance and care must be taken in its measurement if misleading results are to be avoided. It is hoped that these possible sources of error will become apparent from the text so that the potential user of OTFs may be on his guard.

Also it is not *always* possible to distinguish the better of two lenses for a particular application from their respective OTF curves, especially if the eye is to be the final imaging system. The eye is such a complex device, being basically a rather poor optical imaging system followed by very sophisticated data-processing in the retina and brain, that its response to the quality of various images is difficult to determine. Consider, for example, two optical systems which are identical except that one produces very clear, sharp images and the other rather less sharp or 'softer' images. If these systems are used to view, say, geometrical patterns then the majority of viewers will prefer the first. If now the patterns are replaced by, say, a human face the preference will be for the second system, which softens the features and does not pick out every blemish in minute detail. OTFs of the human visual system under certain conditions have been obtained however (see, for example, Campbell and Gubisch (1966)), and no doubt further research will improve the situation.

Another point which it is worth making at the outset, although it will soon become apparent, is that the OTF does not provide a single figure of merit for an optical system. In fact to describe a system with any pretence to completeness, series of OTFs at various spatial frequencies, image orientations and field positions are required.

However, despite these limitations the OTF still provides the most generally useful description of optical image quality.

The treatment of the OTF given in this monograph considers only objects illuminated incoherently (the requirements for this will be given in Chapter 2), and this restriction requires some justification. In principle OTFs can be obtained for all states of illumination; coherent, partially coherent and incoherent.

In the coherent case it is found that the transfer function may be identified with the 'pupil function' of the imaging lens, which is composed of the lens wavefront aberration and the spatial transmission characteristic of the physical pupil of the lens. Modification of the latter allows great control over the transfer function of the lens, and this has led to the use of such systems for spatial filtering and optical data-processing. Since the optical requirements of these lenses in terms of aperture and field are not demanding and any aberration content of the pupil function affects the transfer function in an uncontrollable way, these lenses are generally diffraction-limited, and as such have a very simple transfer function. Another area in which coherent illumination may be encountered is microscopy, but again microscope objectives are generally very highly corrected. For these reasons the specification or testing of such systems by means of an OTF is not particularly useful.

Partially coherent illumination is encountered in the vast majority of optical systems. Its mathematical treatment is rather complicated, and the difficulty of computing or measuring a transfer function with partially coherent light lies mainly in producing a specific degree of partial coherence in the illumination of the object. However, the effects of partial coherence may often not be of any consequence to the overall measurement of an OTF. For example consider the cascaded imaging system shown in Fig. 1 (*a*). The object is illuminated incoherently, and lens A has a spherical aberration which is equal and opposite in sign to that of lens B, as is shown by the different primary images I_F and I_P formed by full aperture and paraxial rays respectively after traversing lens A, and a single perfect image I after imaging by lens B. Axial OTFs of the two lenses separately might give the curves shown in Figs. 1 (*b*) and (*c*). Now to obtain the total effect of the two lenses one might be tempted simply to multiply the two OTF curves, but this obviously cannot give the true, perfect result shown in Fig. 1 (*d*). The reason for the discrepancy is that the aberration of lens A causes the image presented to lens B to be partially coherent so that measurement of its OTF in the normal way (i.e. with incoherent illumination) is not valid. This is also true if the intermediate image is virtual, so that a complex lens consisting of several elements may contain several such partially coherent intermediate images. However, provided that the lens is treated as a whole, or, in the example above, if the OTF is measured from the object to the final image I, the intermediate partial coherence effects may be ignored.

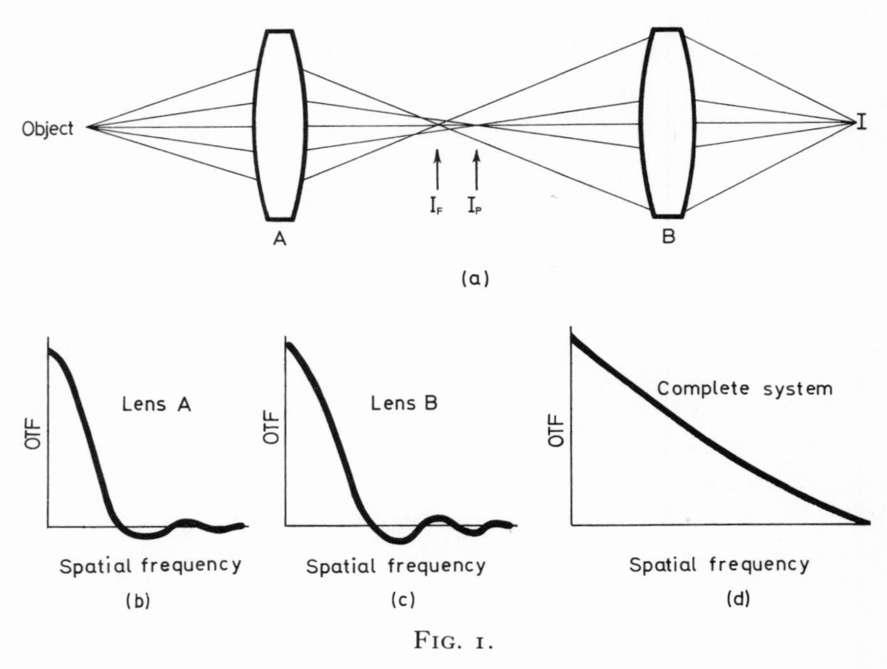

FIG. 1.

As a rider to this, if the intermediate images are of sufficiently high quality then they may be considered to be incoherently illuminated, and the product of the OTFs then does give a true overall transfer function.

A major source of partial coherence which cannot be ignored is illumination of an object by an insufficiently well-corrected condenser; consideration of such systems in microscopy led Hopkins (1951, 1953) to study the theory of partial coherence. In practice the effects of partial coherence on photographic, TV and similar systems may generally be ignored; this is borne out by the close agreement between calculation and measurement based upon incoherent illumination.

The reader is referred to Born and Wolf (1965), Thompson (1969), and Martin (1966), as well as Hopkins, for more detailed treatments of coherent and partially coherent cases.

2

Definition and Formulation of the Polychromatic Optical Transfer Function

2.1. FOUNDATIONS OF OTF THEORY

The concept of the OTF is based on the fact that under certain conditions an optical imaging system can be considered as a passive linear filter with respect to two-dimensional spatial frequencies present in the object, provided that two conditions are satisfied.

The first of these conditions, that of *stationarity,* requires that if the object is translated in the object plane, then the image is correspondingly translated without any other change. This means that the wavefront aberration produced by the imaging lens must be independent of the position of the object, which, except with a few very highly corrected lenses, is not the case over large fields. However, provided that the object is restricted to a region of the object plane over which the wavefront aberration produced by the lens is sufficiently constant, this condition can be tolerably well satisfied. This condition is also known as the condition of *isoplanatism,* and is discussed at length by Dumontet (1955).

The second condition, that of *linear superposition,* requires that the quantity used to describe the effect of an object at a point in the image plane is such that the total effect at the point produced by two or more objects is given by the linear summation of the individual values of the quantity at the image point for each object. This condition may be satisfied for the majority of optical imaging systems, but the quantity in which the system is linear depends on the state of coherence of the object illumination. If the object is illuminated incoherently (or is self-luminous), then the optical system is linear in intensity; if the object is illuminated coherently, then it is linear in complex amplitude; and if the object is illuminated by partially coherent light then consideration of the mutual intensity allows the derivation of a transfer function. For a discussion of this see, for example, Born and Wolf (1965), §10.5.3.

Optical transfer functions can (in principle) be obtained for all states of illumination, but for the reasons outlined in the introduction only incoherent illumination will be considered here, since this case is of the most practical interest. For the illumination of an object to be incoherent it must be both spatially and temporally incoherent.

Spatial incoherence implies that the phases of the monochromatic components of the light from all points of the object are changing at random. This is automatically assured if the object is an extended thermal source, since the wave-trains emitted by the separate atoms of the source are statistically random and therefore incoherent. However, if the object is illuminated by means of a condenser it is necessary to ensure that the region over which the point spread-function of the condenser at the object has a significant intensity is small compared with the finest detail of interest in the object. (The point spread function is the intensity distribution in the image of a point source and is dependent on the aperture of the imaging system, its aberrations and also upon diffraction effects.) If this is not the case the object will be illuminated by partially coherent light. See, for example, Born and Wolf (1965), §10.5.1.

Temporal incoherence of the illumination of an object is implicit in the measurement of monochromatic and polychromatic OTFs, but is generally not explicitly discussed. In fact, for thermal sources and the response times of detectors at present available, the illumination of the object is effectively incoherent (see Appendix 1). Thus the total (time-averaged) intensity Q produced at a point by a thermal source is given by the sum of the monochromatic components; that is,

$$Q = \int_0^\infty Q_\nu(\nu)\,\mathrm{d}\nu$$

where ν is the temporal frequency of the light. This result and its equivalent in terms of wavelength λ, namely,

$$Q = \int_0^\infty Q_\lambda(\lambda)\,\mathrm{d}\lambda \tag{2.1}$$

are of the greatest importance in discussing the polychromatic OTF.

2.2. THE MONOCHROMATIC OTF

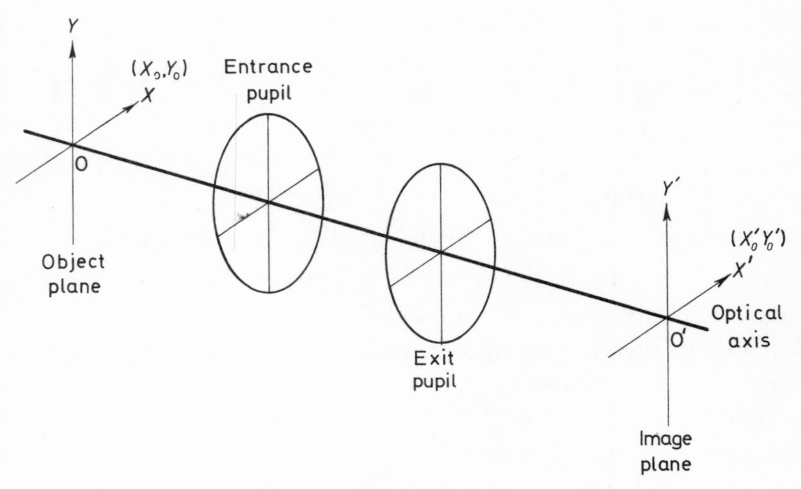

FIG. 2. Basic coordinate system.

Consider an object intersecting the axis of an imaging system at O as in Fig. 2. Cartesian axes X, Y and X', Y' are constructed in the object and image planes. Let the Gaussian image of point (X_O, Y_O) in the object plane be (X'_O, Y'_O) and the transverse magnification be m. Then

$$X'_O = mX'_O \quad \text{and} \quad Y'_O = mY_O$$

For convenience this magnification is taken into account by converting to dimensionless coordinates x, y such that

$$\frac{X_O}{d} = \frac{X'_O}{md} = x \quad \text{and} \quad \frac{Y_O}{d} = \frac{Y'_O}{md} = y$$

where d is the unit of length.

Quantities and coordinates referring to the image space will be primed thus: x'. Monochromatic quantities will carry a subscript λ.

Now each point of the object will give rise to an intensity distribution in the image of the form

$$P_\lambda(x'-x, y'-y)$$

which is called the point spread function. This will be the same for all object points, from the assumption of stationarity (isoplanatism). Because of the assumption of linear superposition the intensity distribution Q'_λ in the image will be given by the convolution

$$Q'_\lambda(x', y') = \iint_{-\infty}^{\infty} P_\lambda(x'-x, y'-y) \cdot Q_\lambda(x, y)\, \mathrm{d}x\mathrm{d}y \qquad (2.2)$$

(see, for example, Born and Wolf (1965), § 9.5.2). The double integration strictly speaking is over the object only, but it may be formally extended to infinity provided that the rest of the object space is not illuminated.

The Fourier transform of Q_λ may be defined as

$$q_\lambda(f, g) = \iint_{-\infty}^{\infty} Q_\lambda(x, y)\, \mathrm{e}^{2\pi \mathrm{i}(fx+gy)}\, \mathrm{d}x\mathrm{d}y \qquad (2.3)$$

where f and g are dimensionless spatial frequencies given by

$$f = \frac{1}{x_c} = \frac{md}{X'_{O,c}} = \frac{d}{X_{O,c}}$$

$$g = \frac{1}{y_c} = \frac{md}{Y'_{O,c}} = \frac{d}{Y_{O,c}}$$

Here x_c etc. are the projected lengths of one cycle of the spatial frequency.

Similar transforms of Q'_λ and P_λ give corresponding expressions for q'_λ and p_λ.

The actual spatial frequencies F_O, G_O and F_I, G_I relating to the object and image spaces respectively are given by

$$F_O = \frac{1}{X_{O,c}} = \frac{f}{d} \qquad F_I = \frac{1}{X'_{O,c}} = \frac{f}{md}$$

$$G_O = \frac{1}{Y_{O,c}} = \frac{g}{d} \qquad G_I = \frac{1}{Y'_{O,c}} = \frac{g}{md}$$

Then, by the convolution theorem, equation (2.2) gives

$$q'_\lambda(f, g) = p_\lambda(f, g) \cdot q_\lambda(f, g)$$

or

$$p_\lambda(f, g) = \frac{q'_\lambda(f, g)}{q_\lambda(f, g)} \qquad (2.4)$$

and this equation defines the monochromatic optical transfer function $p_\lambda(f, g)$ which will, in general, be a complex number. The OTF may be normalized to unity at zero spatial frequencies, and the normalized monochromatic OTF $\bar{p}_\lambda$ is given by

$$\bar{p}_\lambda(f, g) = \frac{p_\lambda(f, g)}{p_\lambda(0, 0)}$$

Thus the optical transfer function is defined as the ratio of the spatial frequency spectrum of the image to that of the object, and is the Fourier transform of the point spread function. Equation (2.4) also shows the usefulness of the OTF in specifying the performance of a lens, since if the OTF and the object intensity distribution are known it is possible, in principle, to calculate the image intensity distribution exactly.

2.3. THE POLYCHROMATIC OTF

The discussion of the monochromatic OTF in the previous section can be extended to polychromatic light in two ways; by consideration of the polychromatic point spread function or by the summation of monochromatic OTF values.

2.3.1. *The polychromatic OTF from the polychromatic point spread function*

From equations (2.1) and (2.2) the total polychromatic intensity distribution in the image of an object of intensity distribution $Q(x, y)$ will be given by

$$Q'(x', y') = \int_0^\infty Q'_\lambda(x', y')\,d\lambda$$

$$= \int_0^\infty R_\lambda . S_\lambda . T_\lambda . \iint_{-\infty}^{\infty} P_\lambda(x'-x, y'-y) . Q_\lambda(x, y)\,dx\,dy\,d\lambda$$

where S_λ is the normalized spectral energy distribution of the source

T_λ is the normalized spectral transmission of the optical system

R_λ is the normalized spectral response of the detector.

Provided that the intensity distribution of the object is independent of wavelength, this can be written:

$$Q'(x',y') = \iint_{-\infty}^{\infty} Q(x,y) \int_0^{\infty} R_\lambda . S_\lambda . T_\lambda . P_\lambda(x'-x,y'-y)\,\mathrm{d}\lambda . \mathrm{d}x\mathrm{d}y$$

$$= \iint_{-\infty}^{\infty} Q(x,y) . P(x'-x,y'-y)\,\mathrm{d}x\mathrm{d}y \qquad (2.5)$$

where

$$P(x',y') = \int_0^{\infty} R_\lambda . S_\lambda . T_\lambda . P_\lambda(x',y')\,\mathrm{d}\lambda \qquad (2.6)$$

and is the polychromatic point spread function.

Now, putting P, Q and Q' in terms of their Fourier transforms p, q and q' as before (c.f. equation (2.3)) we obtain by the convolution theorem:

$$q'(f,g) = p(f,g) . q(f,g) \qquad (2.7)$$

or

$$p(f,g) = \frac{q'(f,g)}{q(f,g)} \qquad (2.8)$$

Comparing this with equation (2.4) shows that the polychromatic OTF is the Fourier transform of the polychromatic point spread function, namely $p(f,g)$. The normalized polychromatic OTF is given by

$$\bar{p}(f,g) = \frac{p(f,g)}{p(0,0)}$$

2.3.2. *The polychromatic OTF from monochromatic OTF values*

Taking the Fourier transform of both sides of equation (2.6),

$$p(f,g) = \iint_{-\infty}^{\infty} e^{2\pi i(fx'+gy')} \int_0^{\infty} R_\lambda . S_\lambda . T_\lambda . P_\lambda(x',y')\,\mathrm{d}\lambda . \mathrm{d}x'\mathrm{d}y'$$

and changing the order of integration gives

$$p(f,g) = \int_0^{\infty} R_\lambda . S_\lambda . T_\lambda . \iint_{-\infty}^{\infty} P_\lambda(x',y') e^{2\pi i(fx'+gy')}\,\mathrm{d}x'\mathrm{d}y' . \mathrm{d}\lambda$$

$$= \int_0^{\infty} R_\lambda . S_\lambda . T_\lambda . p_\lambda(f,g)\,\mathrm{d}\lambda \qquad (2.9)$$

Hence, the polychromatic OTF can be obtained from a weighted sum of the monochromatic OTF values.

There are thus basically two approaches to the determination of polychromatic OTFs. Either the ratio of the spatial spectra of image and object can be found for polychromatic light, or the sum of the suitably weighted monochromatic OTFs can be found.

2.4. DISCUSSION

To understand further the implications of equations (2.4) and (2.8) and the physical meaning of the OTF, it is perhaps simplest to consider the result of imaging an object composed of a single spatial frequency, that is, an illuminated trans-

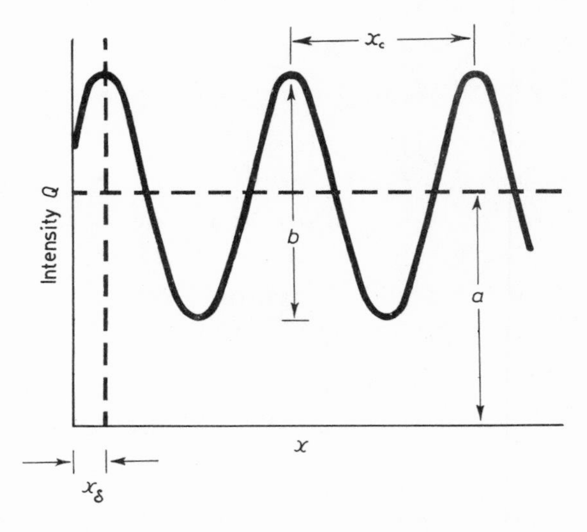

FIG. 3. Co-sinusoidal object.

parency with a transmission varying co-sinusoidally with x (see Fig. 3), thus:

$$Q(x, y) = a + b\cos(2\pi f_c x + \delta) \tag{2.10}$$

where a is the mean intensity level

b is the peak-to-peak modulation, and must be equal to or less than a

δ is the spatial phase angle corresponding to x_δ, the position of the first maximum of transmission from the origin

f_c is the spatial frequency $1/x_c$, x_c being the length of one cycle.

The contrast of the object is defined as b/a.

In what follows polychromatic illumination will be assumed, but the same reasoning applies to monochromatic illumination.

From equation (2.2) the image intensity is given by

$$Q'(x', y') = \iint_{-\infty}^{\infty} P(x'-x, y'-y) . \{a+b\cos(2\pi f_c x+\delta)\} \, dxdy$$

The integral of the point spread function P over y is, by the assumptions of superposition and stationarity, the image of a line object, and is called the line spread function $L(x')$. Then,

$$Q'(x') = \int_{-\infty}^{\infty} \{a+b\cos(2\pi f_c x+\delta)\}.L(x'-x)\, dx$$

and, making the coordinate transformation $\mathrm{x} = x'-x$, this may equally be written

$$Q'(x') = \int_{-\infty}^{\infty} \{a+b\cos\left(2\pi f_c(x'-\mathrm{x})+\delta\right)\}.L(\mathrm{x})\, d\mathrm{x}$$

$$= a \int_{-\infty}^{\infty} L(\mathrm{x})\, d\mathrm{x}$$

$$+ b\cos(2\pi f_c x'+\delta) . \int_{-\infty}^{\infty} L(\mathrm{x}) \cos 2\pi f_c \mathrm{x} \, d\mathrm{x}$$

$$+ b\sin(2\pi f_c x'+\delta) . \int_{-\infty}^{\infty} L(\mathrm{x}) \sin 2\pi f_c \mathrm{x}\, d\mathrm{x}$$

The last two integrals in this equation are the cosine and sine transforms of the line spread function and will be written as $C(f_c)$ and $S(f_c)$ respectively. If now the unit of intensity is scaled to make the integral over x of the line spread function unity, we have:

$$Q'(x') = a + b\cos(2\pi f_c x'+\delta) . C(f_c) + b\sin(2\pi f_c x'+\delta) . S(f_c)$$

$$= a + M(f_c) . b\cos\left(2\pi f_c x'+\delta+\varepsilon(f_c)\right)$$

where

$$M(f_c) = \sqrt{C(f_c)^2+S(f_c)^2}$$

and

$$\varepsilon(f_c) = \tan^{-1}\left(-\frac{S(f_c)}{C(f_c)}\right).$$

Thus the image has the same form as the object (except for a linear magnification), but its contrast is changed by a factor $M(f_c)$, the *contrast transfer function*, and its spatial phase is shifted by $\varepsilon(f_c)$, the *phase transfer function*, which corresponds to a displacement of the image relative to the Gaussian image of the object.

Now consider the complex number

$$M(f_c) \,.\, e^{i\varepsilon(f_c)} = C(f_c) - i\, S(f_c)$$

$$= \int_{-\infty}^{\infty} L(x) \,.\, e^{2\pi i f_c x}\, dx \qquad (2.11)$$

This is the one-dimensional Fourier transform of the line spread function, and as stated previously the line spread function is the integral of the point spread function over one dimension; that is,

$$L(x') = \int_{-\infty}^{\infty} P(x'-x, y'-y)\, dy'$$

Taking the one-dimensional Fourier transform of this equation gives

$$\int_{-\infty}^{\infty} L(x')\, e^{2\pi i f_c x'}\, dx' = \iint_{-\infty}^{\infty} P(x'-x, y'-y)\, e^{2\pi i f_c x'}\, dx' dy'$$

$$= p(f_c, 0)$$

which is the OTF at spatial frequency f_c. Comparing this with equation (2.11) shows that $M(f_c) . e^{i\varepsilon(f_c)}$ is the OTF at spatial frequency f_c.

Thus, the real part of the OTF gives the contrast transfer function (sometimes called the *modulation transfer function*, MTF), and the imaginary part gives the phase transfer function.

Hence we have

$$M(f) = \left| \int_{-\infty}^{\infty} L(x)\, e^{2\pi i f x}\, dx \right|$$

Using Schwarz's inequality gives

$$M(f) \leqslant \int_{-\infty}^{\infty} \left| L(x)\, e^{2\pi i f x} \right| dx$$

$$\leqslant \int_{-\infty}^{\infty} L(x)\, dx \quad \text{(since } L(x) \text{ is real)}$$

But $\int_{-\infty}^{\infty} L(x)\,dx$ is the value of the OTF at zero spatial frequency, so that the contrast of the image can never be greater than that of the object. It will be shown later in Chapter 3 that in fact the contrast of an image, even when produced by an aberrationless lens, in general decreases with increasing spatial frequency, to an extent dependent on the pupil shape.

Thus if the spatial frequency components of an object are found by Fourier analysis, the OTF describes the contrast and phase of these components in the image. If the point spread function of the imaging system is symmetrical (i.e. an even function) its transform, the OTF, is real and there is no phase shift in the image.

3

Methods of Measurement

3.1. INTRODUCTION

Various forms of equipment have been developed to measure OTFs but they all use one of two basic techniques—*image scanning* and *shear interferometry*.

The scanning methods, in which an image is scanned with a screen having a certain spatial transmission characteristic and the total transmitted light flux recorded, can be further divided into *indirect* and *direct scanning*. In the former, an image is scanned to give its intensity distribution and then the Fourier transform of this is divided by the Fourier transform of the object to give the OTF, as indicated by equations (2.4) or (2.8). As this is essentially a two-stage process the term 'indirect' is used. The direct scanning methods scan the image of a slit with a periodic grating (or vice versa) and the transmitted light flux then gives direct information of the OTF at the spatial frequency of the grating.

The interferometric methods shear the wavefront in the pupil of the lens under test and measure the total light flux passed by the system as the optical path in one arm of the interferometer is continuously varied.

Both the indirect and direct scanning methods can be used to measure polychromatic OTFs. The interferometric methods, although they can be used with polychromatic light, can only be used to measure monochromatic OTFs.

This chapter describes the theory and practice of these methods of measurement. The methods and experimental details are not meant to be complete, but they are representative; they deal mainly with the fundamental principles and those aspects especially relevant to polychromatic OTFs. The reader is referred to the original papers for more complete experimental details.

3.2. SCANNING METHODS

3.2.1. *Theory*

In what follows it will be assumed that polychromatic illumination is used.

Let the image, of intensity distribution $Q'(x', y')$, produced by a lens under test be scanned by a screen having a transmission distribution $Z(x', y')$. Then the light flux Φ passed by the screen when the latter is displaced distances ξ, η along the x', y' axes is given by

$$\Phi(\xi, \eta) = \iint_{-\infty}^{\infty} Q'(x', y') \cdot Z(x' - \xi, y' - \eta)\, \mathrm{d}x' \mathrm{d}y'$$

If Z is an even function then

$$Z(x' - \xi, y' - \eta) \equiv Z(\xi - x', \eta - y')$$

and the integral of the previous equation becomes a convolution. Taking the Fourier transform of both sides of this equation then gives, by the convolution theorem,

$$\phi(f, g) = q'(f, g) \cdot z(f, g)$$

where ϕ and z are the Fourier transforms of Φ and Z respectively. Substituting for $q'(f, g)$ from equation (2.7),

$$\phi(f, g) = p(f, g) \cdot q(f, g) \cdot z(f, g) \tag{3.1}$$

An important point should be noted here. This result indicates that it makes no difference whether Q is the object and Z the screen or vice versa.

3.2.2. *Indirect methods*

In these methods, objects are chosen which either have simple spatial spectra or are easily manufactured. The simplest object is a single point, which may be represented mathematically by a delta function, $\delta(x, y)$. The transform of a delta function is unity so that all spatial frequencies are present to the same extent in the object. If the scanning screen is also a single point aperture we have:

$$Q(x, y) = \delta(x, y) \qquad q(f, g) = 1$$
$$Z(x', y') = \delta(x', y') \qquad z(f, g) = 1$$

and equation (3.1) becomes

$$p(f, g) = \phi(f, g)$$

This method was used by Naish and Jones (1954) to give the intensity distribution in the point spread function (see Fig. 4). The lens C adjusts the effective distance of the pinhole object S to be at the working distance of the lens L under

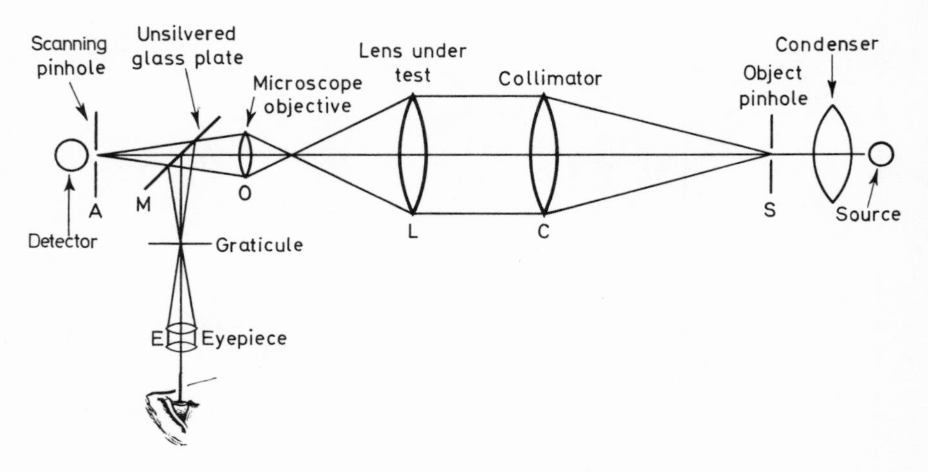

FIG. 4. Apparatus used by Naish and Jones.

test, and the image of S formed by L is magnified by the microscope objective O so that the size of the scanning aperture A is not impossibly small. Scanning is done by moving the pinhole S. The beam-splitter M reflects part of the light onto a graticule and eyepiece at E which enables the coordinates of S relative to A to be determined.

For accurate polychromatic measurements of lens L to be made, both the lens C and microscope objective O must be chromatically well corrected, since the method in effect measures the OTF of the combination of all three.

The great disadvantage of using a point object is the very small amount of light available. This causes inaccuracies through noise effects in the detector, and to avoid this slit objects have been used. As was shown in Chapter 2, the OTF is the one-dimensional Fourier transform of the line spread function, and equation (3.1) can be written

$$\phi(f, 0) = p(f, 0) \cdot q(f, 0) \cdot z(f, 0)$$

For convenience this will be written

$$\phi(f) = p(f) \cdot q(f) \cdot z(f) \tag{3.2}$$

Again the simplest forms for Q and Z are delta functions, corresponding to a slit object and a parallel slit scanning aperture. Then equation (3.2) becomes

$$p(f) = \phi(f)$$

Ingelstam and Lindberg (1951) used this method to record the line spread function, using a one-micron wide scanning slit. Kuwabara (1955) improved on the method by making

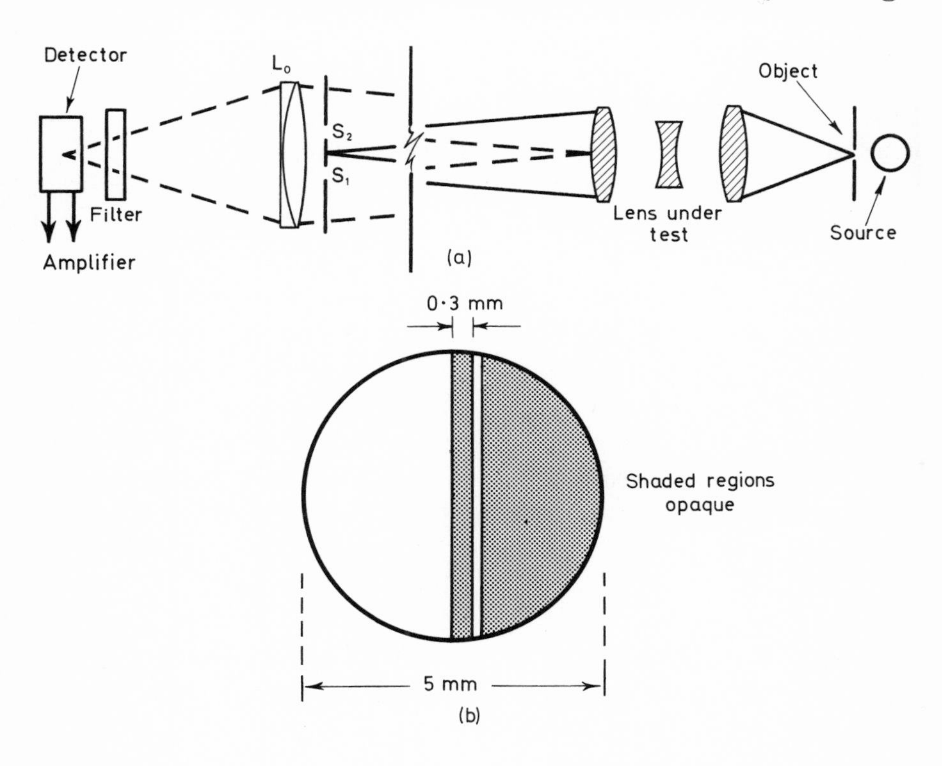

FIG. 5. (*a*) Apparatus and (*b*) object used by Kuwabara.

allowances for slow variations in source intensity. The apparatus he used is shown schematically in Fig. 5 (*a*). The object consisted of a glass disk, one half of which was transparent and the other half opaque except for a slit of width 1·7 micron running parallel to the straight dividing edge and separated from it by 0·3 mm, this distance being considered sufficient for there to be no overlap of the edge and slit images. The object is shown in Fig. 5 (*b*). Of the two slits in the image plane of the lens under test, S_1 was movable and S_2 fixed, with S_2 lying on the bright side of the image, well away from the boundary. Each slit had a shutter. Lens L_O acts as a field lens and, by ensuring that light through either S_1 or S_2 falls on the same part of the detector, prevents errors due to non-uniform sensitivity of the detector surface. By moving the slit S_1 across the image and opening the shutters of S_1 and S_2 alternately, the ratios of the intensities in the slit image to the intensity in the bright part of the image were obtained, and the line spread function traced.

Kuwabara in fact used monochromatic light from a Mercury lamp and a No. 77 Wratten filter, but the apparatus could be used as it stands with a polychromatic source. The field lens L_O need not be well corrected chromatically since the measurement is one of comparison and both channels will be affected equally.

All the methods so far considered have produced a curve showing the intensity distribution in the point or line spread function, and to obtain the OTF it is then necessary to compute the Fourier transform of these. To remove the necessity for this calculation, the line spread function may be scanned in such a way as to produce a time-varying electrical signal having the same form as the intensity distribution; this signal is then analysed into its component frequency spectrum by conventional electronic methods, to give the spatial frequency spectrum of the image.

One way of doing this is to scan the image of a slit with a train of identical equally spaced slits travelling at a velocity V in a direction perpendicular to their length, the separation l of the slits being much greater than the width of the line spread function. The light flux passed by the system then takes the form of a train of pulses as shown in Fig. 6, and this may be represented as the convolution of the image intensity

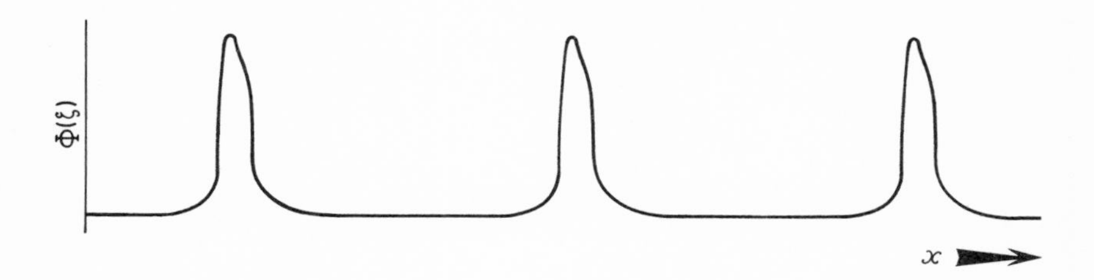

FIG. 6. Pulse train obtained by multiple slit scanning.

distribution with an infinite array of equi-spaced delta functions, thus:

$$\Phi(\xi) = \int_{-\infty}^{\infty} Q'(x') \cdot \sum_{k=-\infty}^{\infty} \delta(\xi + kl - x')\, \mathrm{d}x'$$

where k is an integer.

By the convolution theorem this gives

$$\phi(f) = q'(f) \cdot \sum_{k=-\infty}^{\infty} \frac{1}{|l|} \cdot \delta\left(f - \frac{k}{l}\right)$$

$$= \frac{1}{|l|} \sum_{k=-\infty}^{\infty} q'\left(\frac{k}{l}\right)$$

Thus ϕ has non-zero value only at $f = k/l$, and we may write

$$\sum_{k=-\infty}^{\infty} \phi \frac{k}{l} = \frac{1}{|l|} \sum_{k=-\infty}^{\infty} q'\left(\frac{k}{l}\right)$$

Substituting for q' in equation (2.7), and recalling that $q = 1$ for a slit object, we find

$$\sum_{k=-\infty}^{\infty} \phi\left(\frac{k}{l}\right) = \frac{1}{|l|} \sum_{k=-\infty}^{\infty} p\left(\frac{k}{l}\right)$$

Now a spatial frequency f being scanned at a velocity V by a slit gives rise to a signal in the detector which oscillates at a temporal frequency ν given by:

$$\nu = V.f$$

Thus an output I from the detector is obtained, of the form

$$I(v) = \rho \sum_{k=-\infty}^{\infty} \phi(kv_l)$$

where ρ is the sensitivity of the detector, and v_l is the temporal frequency corresponding to the fundamental spatial frequency $1/l$.

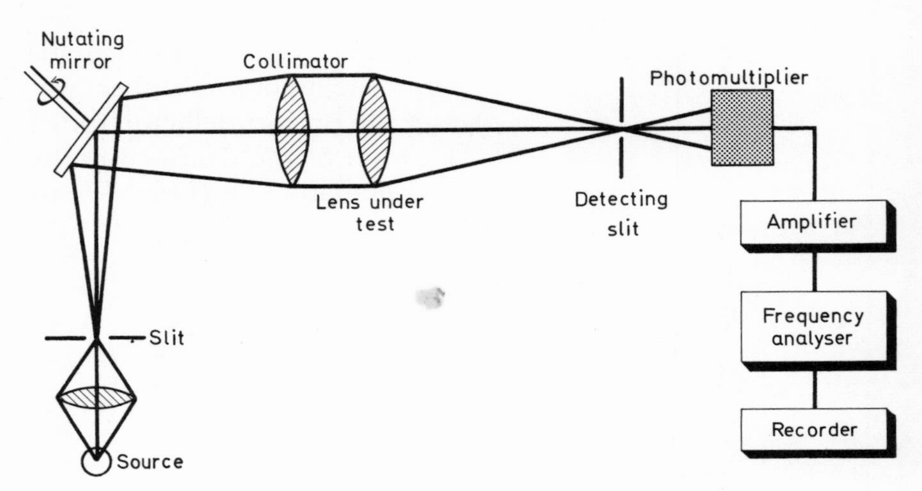

FIG. 7. Apparatus used by Polster.

Then, if the signal from the detector is passed to a narrow-bandwidth detector tuned to accept frequency jv_l only, where j is a specified integer, the amplitude and phase of the detected signal correspond directly to the amplitude and phase of the OTF at the spatial frequency given by

$$f = \frac{jv_l}{V} \tag{3.3}$$

Thus, either the scanning velocity may be kept constant, in which case it is necessary to have many narrow-bandwidth detectors to obtain measurements at various spatial frequencies, or the scanning velocity may be made variable, in which case it is only necessary to have one narrow-bandwidth detector.

Polster (1955) employed the former method, using a frequency analyser as a variable narrow-bandwidth detector. Instead of using a train of slits he arranged the image of the object slit to move in an elliptical locus by means of a nutating mirror rotating at constant speed (see Fig. 7). The scanning slit was fixed, and was situated at the end of one of the minor axes of the ellipse so that the image was scanned once per revolution. The system can be used with polychromatic light provided that the auxiliary lens (collimator) is sufficiently well corrected chromatically.

Birch (1961) scanned the image of a slit with a drum carrying 24 identical slits. His apparatus is shown schematically in Fig. 8. The signal from the photomultiplier was passed to a narrow-bandwidth amplifier and detector tuned to 1000 Hz, and the speed of rotation of the drum could be adjusted to satisfy the condition of equation (3.3) for various

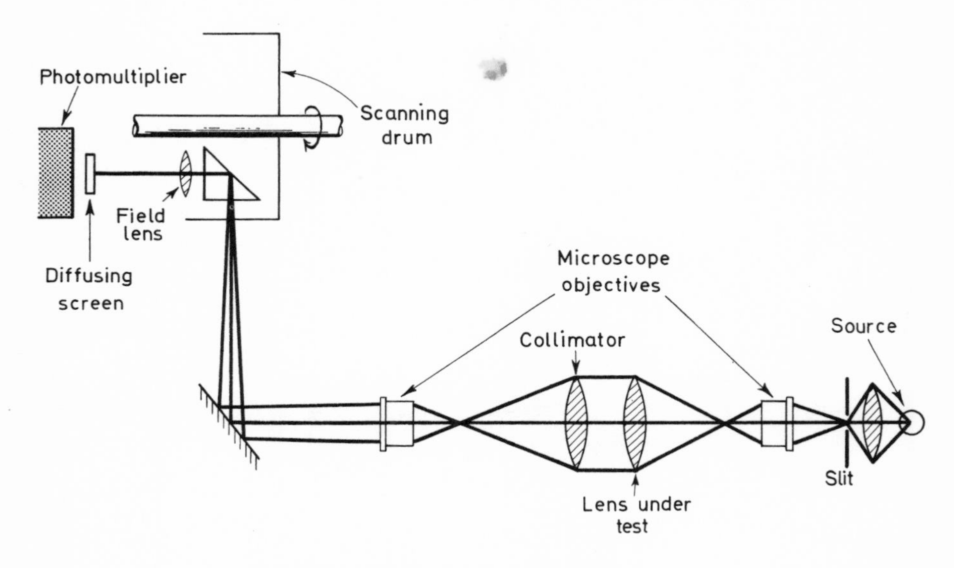

FIG. 8. Apparatus used by Birch.

values of j. No attempt was made at this time to measure the phase of the transfer function, but later Birch compared the phases of the harmonics with those of the same frequency from another, symmetrical slit image incident on the drum, which because of its symmetry had no phase error.

It should be noted that three auxiliary lenses were used by Birch: a microscope objective to reduce the size of the effective object slit so that a standard spectrometer slit could be used; a collimating lens following the lens under test; and another microscope objective to transfer the slit image onto the scanning drum. All of these must be chromatically corrected for polychromatic OTFs to be measured with any accuracy.

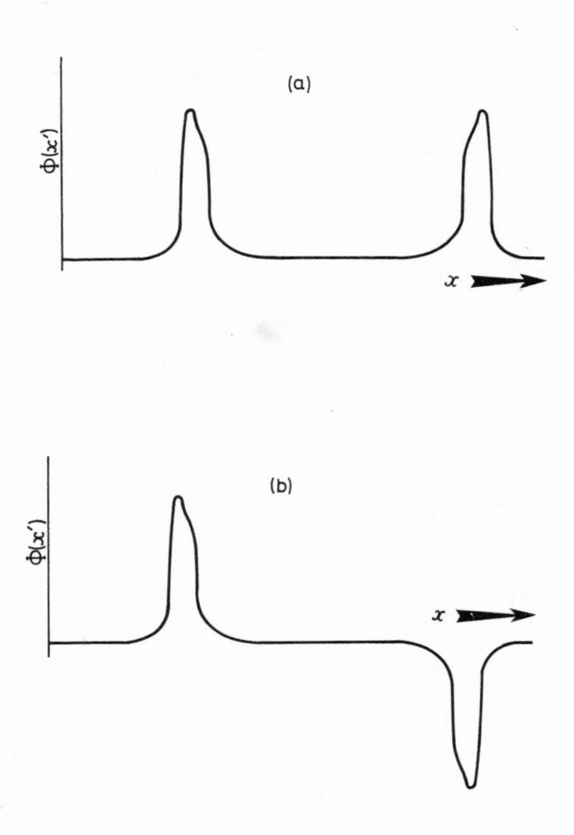

FIG. 9. (*a*) Pulse train obtained by scanning with a reciprocating slit. (*b*) As (*a*), but with alternate pulse electrically inverted.

Ose and Takashima (1965) refined the slit scanning technique to give more accurate phase information. If the image of a narrow slit is scanned backwards and forwards by another narrow slit travelling at constant velocity V during a traverse, then the electrical signal from a detector measuring the transmitted flux is a series of pulses as shown in Fig. 9(*a*). These pulses correspond to $Q'(x')$ and $Q'(-x')$ alternately. Taking the turning point between two pulses as the x' origin, the pulse train may be represented by a Fourier series, thus:

$$\Phi(x') = \Phi_0 - S_1 \cos 2\pi . fx' - C_2 \cos 2\pi . 2fx' + S_3 \cos 2\pi . 3fx' + \dots \quad (3.4)$$

and if alternate pulses are electrically inverted, as in Fig. 9(*b*), the Fourier series representing this is

$$\Phi(x') = C_1 \sin 2\pi . fx' - S_2 \sin 2\pi . 2fx' - C_3 \sin 2\pi . 3fx' + \dots \quad (3.5)$$

where $\Phi_0 = \frac{1}{|2l|} \int_{-l}^{l} Q'(x')\, dx'$

$$C_k = \frac{1}{|2l|} \int_{-l}^{l} Q'(x') \cos 2\pi kf . 2l\, dx' \quad (3.6)$$

$$S_k = \frac{1}{|2l|} \int_{-l}^{l} Q'(x') \sin 2\pi kf . 2l\, dx' \quad (3.7)$$

k being an integer.

Provided that the distance $2l$ is sufficiently large, so that $Q'(x')$ falls to zero within it, the limits of these integrals may be formally taken as $\pm\infty$. As stated above, the object is a narrow slit, which may be considered as equivalent to a line object, so that $Q'(x')$ is the line spread function and the integrals in equations (3.6) and (3.7) are its cosine and sine transforms at spatial frequency $k/2l$. Thus if the time-varying signals corresponding to equations (3.4) and (3.5), namely,

$$I(v) = I_0 - S_1 \cos 2\pi . vt - C_2 \cos 2\pi . 2vt + \dots$$

and

$$I(v) = C_1 \sin 2\pi . vt - S_2 \sin 2\pi . 2vt - \dots$$

are passed to a detector tuned to a frequency of kv, one obtains values of C_k and S_k, and as was shown in Chapter 2 the OTF is given by

$$p\left(\frac{k}{2l}\right) = M\left(\frac{k}{2l}\right) . e^{i\varepsilon(k/2l)}$$

where

$$M\left(\frac{k}{2l}\right) = \sqrt{C_k^2 + S_k^2}$$

and

$$\varepsilon\left(\frac{k}{2l}\right) = \tan^{-1}\left(-\frac{S_k}{C_k}\right)$$

Thus one may obtain a series of measurements of the OTF at spatial frequencies separated by $1/2l$.

Ose and Takashima used a vidicon tube to scan the slit image, but in principle any photo-electric detector could be used.

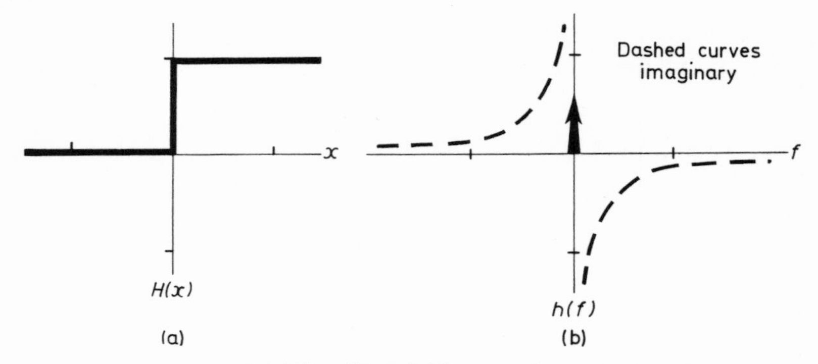

FIG. 10. (*a*) The Heaviside step function, and (*b*) its Fourier transform.

All of these slit scanning methods have two short-comings. The amount of light available, whilst greater than that from point objects, is still small, and detector noise can still be a problem. Also, neither the variable nor the constant velocity scanning methods allow measurement of the OTF at zero spatial frequency, since in the former case an infinite scan velocity would be required, and in the latter case it would be necessary for the set of filters and detectors to have uniform characteristics down to zero temporal frequency, which, whilst being possible in principle, is extremely difficult to attain in practice. This means that the OTF cannot be normalized.

To lessen the problem of low intensity the image of an edge may be scanned with a narrow slit screen, the slit running parallel to the edge. The edge may be represented by the Heaviside step function [Fig. 10(*a*)]:

$$\begin{aligned} H(x) &= 0 \quad \text{for } x < 0 \\ &= \tfrac{1}{2} \quad \text{for } x = 0 \\ &= 1 \quad \text{for } x > 0 \end{aligned}$$

Equation (3.2) then becomes

$$\phi(f) = p(f) \,.\, h(f) \,.\, z(f)$$

where $h(f)$ is the one-dimensional Fourier transform of $H(x)$, shown in Fig. 10(b), i.e.,

$$h(f) = \tfrac{1}{2}\left(\delta(f) - \frac{i}{\pi f}\right)$$

and

$$z(f) = 1$$

Thus we have

$$\phi(f) = p(f).\tfrac{1}{2}\left(\delta(f) - \frac{i}{\pi f}\right)$$

or

$$p(0) = 2\phi(0)$$

$$p(f) = 2i.\phi(f).f\pi \quad \text{for } f \neq 0$$

This method has the disadvantage that the amplitude of the spatial frequency components of the object decrease with increasing spatial frequency, so that the sensitivity of these measurements is reduced.

Another method of improving the photometric efficiency of indirect OTF measurement has been used by Kubota and Ohzu (1957). A screen with randomly spaced regions of transmission is the object, and the image is scanned by a similar screen. As has already been established (equation (2.5)), the object, image and point spread function are related as follows:

$$Q'(x', y') = \iint_{-\infty}^{\infty} P(x'-x, y'-y).Q(x, y)\,dxdy \tag{3.8}$$

The cross-correlation function of the object and image over an area A is given by

$$\Gamma_{1,0}(\xi, \eta) = \frac{1}{A}\iint_A Q(x, y).Q'(x+\xi, y+\eta)\,dxdy$$

Substituting for Q' from equation (3.8) gives

$$\Gamma_{1,0}(\xi, \eta) = \frac{1}{A}\iint_A Q(x, y)\iint_{-\infty}^{\infty} P(x', y')$$

$$\times Q(x-x'+\xi, y-y'+\eta)\,dx'dy'.dxdy$$

If now Q is a statistically random function the order of integration can be reversed, giving

$$\Gamma_{1,0}(\xi, \eta) = \iint_{-\infty}^{\infty} P(x', y')\,\Gamma_{0,0}(\xi-x', \eta-y')\,dx'dy' \tag{3.9}$$

where $\Gamma_{O,O}$ is the autocorrelation function of the object.

Taking the transform of equation (3.9) gives, by the convolution theorem,

$$\gamma_{I,O}(f, g) = p(f, g) \cdot \gamma_{O,O}(f, g)$$

where $\gamma_{I,O}$ and $\gamma_{O,O}$ are the Fourier transforms of $\Gamma_{I,O}$ and $\Gamma_{O,O}$. For a truly random object $\gamma_{O,O}(f, g)$ is a constant and may be scaled to unity, so that

$$p(f, g) = \gamma_{I,O}(f, g)$$

Kubota and Ohzu used the grain structure of exposed and developed photographic emulsions as random object and screen, and measured the cross correlation by linearly translating one in the plane of the image of the other, recording the light flux passed by the system as a function of screen displacement (see Fig. 11). A Fourier transformation was then performed on the cross-correlation curve to give the OTF.

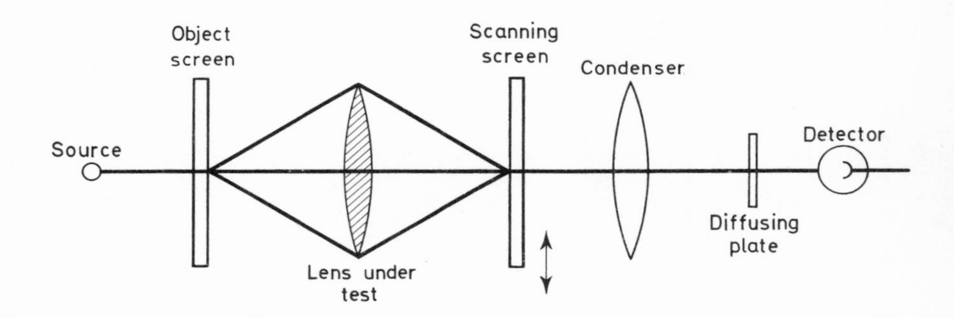

FIG. 11. Apparatus used by Kubota and Ohzu.

In fact, the screens used did not have a sufficiently flat spatial frequency noise spectrum for the autocorrelation function to be taken as unity, and correction had to be made for this.

The method again lends itself to the measurement of the polychromatic OTF, provided that as before any auxiliary lenses are sufficiently chromatically corrected.

3.2.3. *Direct methods*

In these methods the image of a slit is scanned with a screen having a periodic transmission characteristic. This will hereafter be referred to as a grating. Also, as was shown in § 3.2.1, the roles of slit and grating may be reversed, and it is a question of practical convenience as to which arrangement is used.

Gratings having co-sinusoidal transmission versus distance characteristics have been constructed by various workers. Two types are possible. Either the optical density of the grating can be made to vary co-sinusoidally or, since when it is used with a slit object only one dimension is of interest, the grating can be in the form of a co-sinusoidal variation in area of complete transmission in an otherwise opaque screen. The former is known as a *density grating*, and the latter an *area grating*.

The transmission of such gratings is represented by

$$Z(x') = a + b \cos 2\pi f_c x' \quad \text{(c.f. equation 2.10)}$$

where b/a = contrast

f_c = the reciprocal of the length of one period of the grating.

Equation (3.1) then becomes

$$\phi(f) = p(f) . \left[a\delta(f) + \frac{b}{2} \left\{ \delta(f + f_c) + \delta(f - f_c) \right\} \right]$$

Thus

$$\phi(0) = a . p(0)$$

$$\phi(f_c) = \frac{b}{2} . p(-f_c) + \frac{b}{2} . p(+f_c)$$

If either the grating or slit is now moved with constant velocity V, the detector output will be the combination of a zero temporal frequency component and a component at a frequency of ν_c, given by

$$\nu_c = V . f_c \tag{3.10}$$

(positive and negative frequencies being physically identical).

Thus if the detector is adjusted to reject zero frequency, the output from the detector is given by

$$I(\nu_c) = \rho . \phi(\nu_c)$$

where ρ is the sensitivity of the detector. The amplitude and phase of this signal then correspond to the amplitude and phase of the OTF at the spatial frequency f_c given by equation (3.10).

The measurement may then be repeated using gratings of various spatial frequencies to obtain the complete OTF. In practice it is best to limit the bandwidth of the detector to a small region around ν_c to make the signal-to-noise ratio as large as possible.

Measurements of the OTF have been made using density gratings, but the gratings are difficult to produce, relying as they do upon photographic techniques requiring a linear light intensity versus density emulsion characteristic, and they have been superseded by gratings of the area type.

Lindberg (1954) used a set of co-sinusoidal area gratings to measure the modulus of the OTF of a lens, and this method was improved by Ingelstam, Djurle and Sjögren (1956) to include the phase. Their apparatus is shown schematically in Fig. 12. The image of the slit S formed by the lens under test L was made to scan across a screen T at the back focus of collimator C, by means of a motor-driven mirror M. Only one strip of the screen was allowed to pass light to the photomultiplier P at a time, and the output from the photomultiplier was passed to a pen recorder. At the beginning and end of each track on the screen were two phase marks, whose widths were approximately equal to one cycle of one of the highest frequency cosine waves. The peak-to-peak amplitude of the pen recorder trace gave the modulus of the OTF at that particular spatial frequency, and the phase was obtained by measuring the distance between a phase mark and one of the cosine wave peaks and comparing it with the equivalent distance on the screen itself, suitably scaled.

Again this method is quite suitable for the measurement of the polychromatic OTF, provided that the collimator is sufficiently well corrected chromatically. It is however a rather limited technique in that it is only possible to measure the OTF at certain spatial frequencies and the measurement of phase is very tedious. Also it is not possible to obtain the OTF at zero spatial frequency, so the normalized OTF cannot be obtained.

Herriott (1958) used a screen composed of a set of area cosine waves of continuously varying spatial frequency along a length of 35 mm film. This was made to travel at a

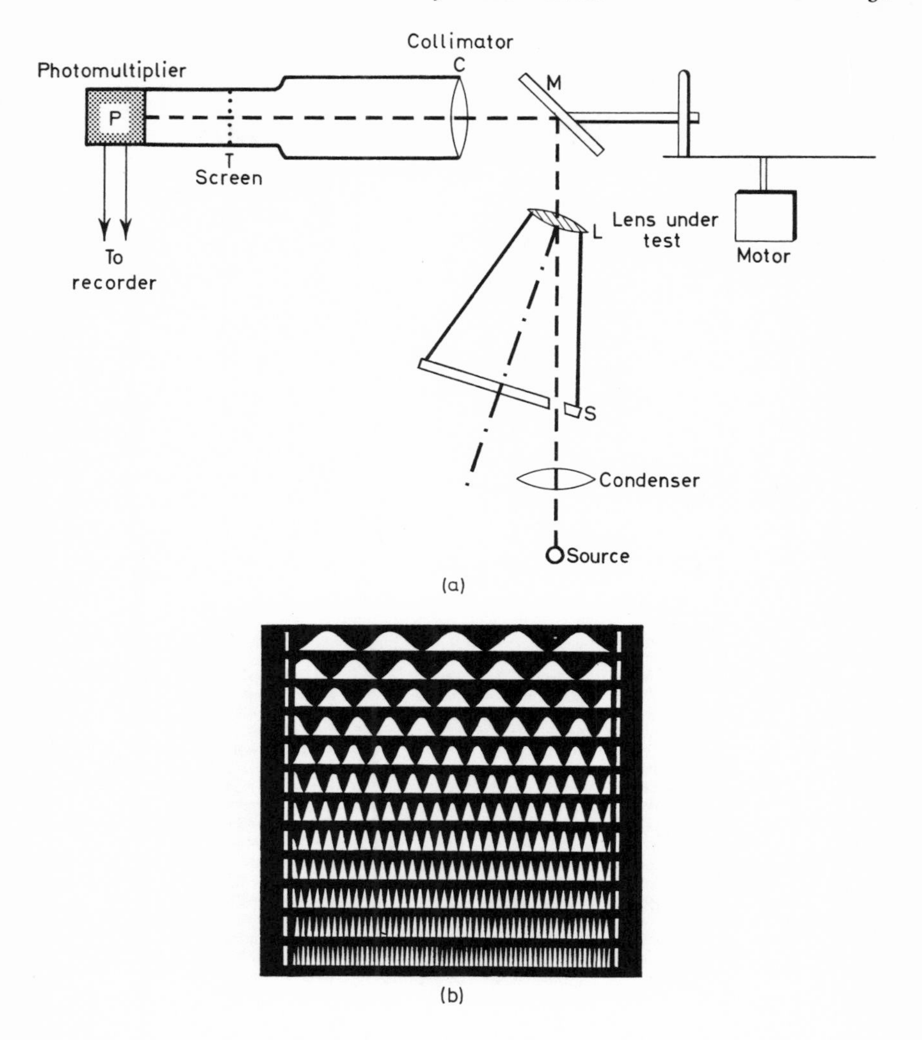

FIG. 12. (*a*) Apparatus, and (*b*) screen used by Ingelstam *et al.* (1956). *(Courtesy, J. opt. Soc. Am.)*

constant velocity and imaged by the lens under test onto a slit followed by a detector. The output from the detector was amplified, rectified and then passed either to the *y*-plates of an oscilloscope or to a pen recorder. The apparatus is shown schematically in Fig. 13. Phase measurement was made by comparing the phase of the signal from the main detector

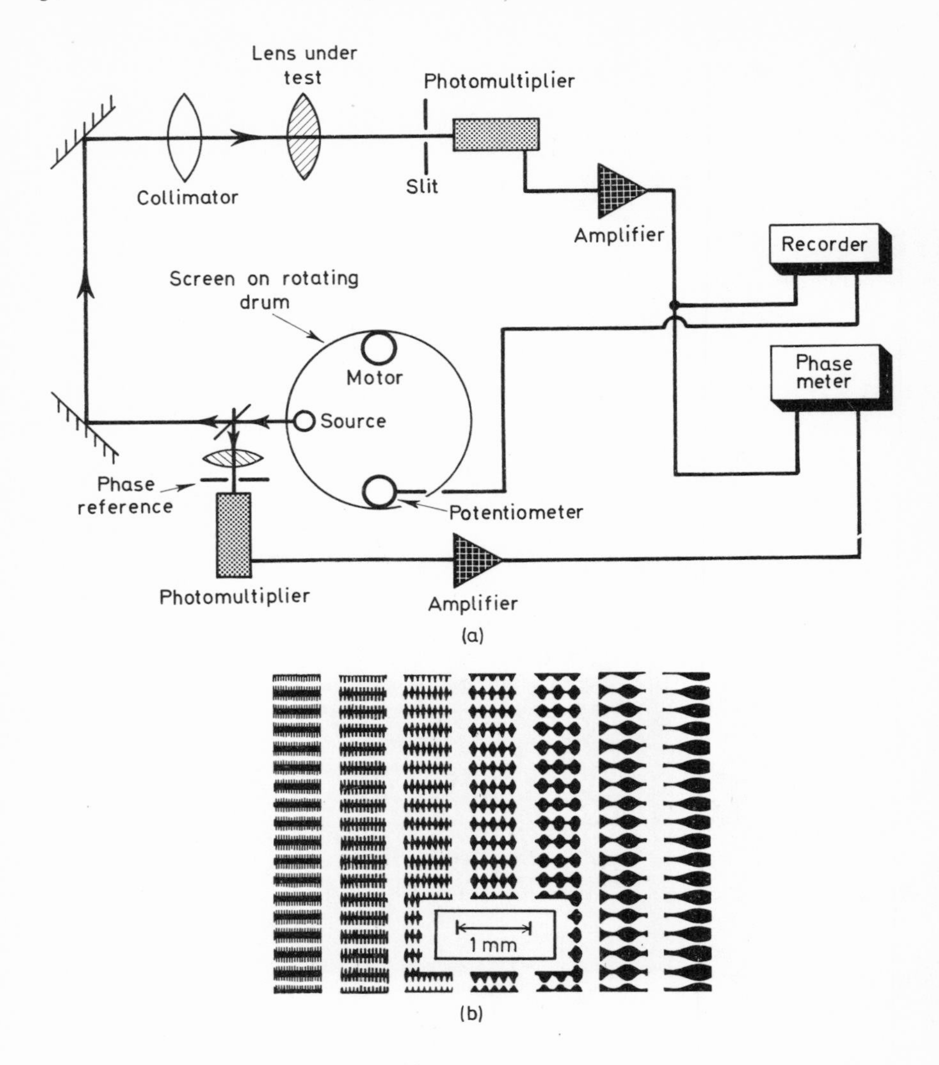

FIG. 13. (*a*) Apparatus, and (*b*) sections of the screen used by Herriott (1958).
(Courtesy, J. opt. Soc. Am.)

with that from a subsidiary detector receiving light from a beam-splitter, lens and slit on the axis as shown in the diagram.

For measurement of the polychromatic OTF the collimating lens must be chromatically corrected. Again this method cannot measure the OTF at zero spatial frequency since this would require the measurement of a d.c. signal, which is technically difficult because of stability problems in d.c. amplifiers. Also gratings of continuously variable spatial frequency are difficult to manufacture.

Since any periodic structure can be resolved into Fourier components, gratings having other than a cosine characteristic may be used, provided that care is taken to ensure that only the contribution of one of the frequency components (usually the fundamental) is accepted by the measuring system. Such gratings are generally more easily and accurately made than cosine gratings.

The transmission of a square-wave grating of unit contrast may be represented by

$$Z(x') = 1 + \frac{4}{\pi}\{\cos(2\pi f_c x') - \tfrac{1}{3}\cos(2\pi . 3f_c x') + \ldots\}$$

where f_c is the reciprocal of the length of one period of the grating. Then for a slit object equation (3.2) becomes

$$\phi(f) = p(f)\Big[\delta(f) + \frac{2}{\pi}\{\delta(f+f_c) + \delta(f-f_c) - \tfrac{1}{3}\big(\delta(f+3f_c) + \delta(f-3f_c)\big) + \ldots\}\Big]$$

If now either the grating or slit is moved with a constant velocity V, the detector output will have a temporal frequency spectrum given by

$$I(\nu) = \rho . \phi(\nu)$$

where ρ is the sensitivity of the detector and

$$\nu = Vf$$

If this signal is passed to a narrow bandwidth filter tuned to exclude all frequencies except ν_c, the amplitude and phase of the output from the filter will be proportional to the amplitude and phase of the OTF at the spatial frequency f_c.

Lindberg (1954) used a square-wave grating in the form of a sector disk. This he caused to rotate at a constant angular velocity about its centre, in the image plane of the lens under test, as shown in Fig. 14. The object used was a pinhole, and the position of the pinhole image relative to the centre of the disk could be varied by moving the disk. This had the effect

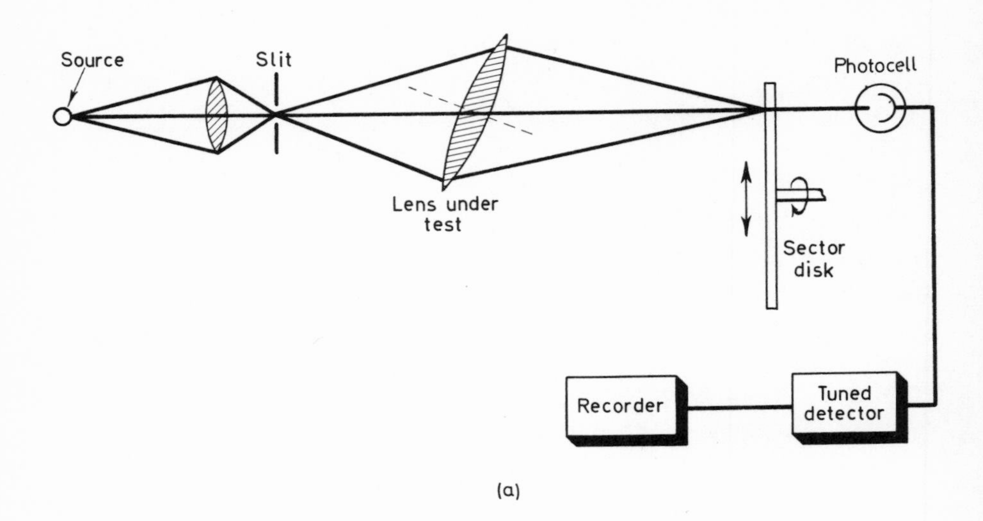

FIG. 14. (*a*) Apparatus, and (*b*) sector screen disk of the type used by Lindberg.

of changing the spatial frequency being measured whilst maintaining the same temporal frequency from the detector. No attempt was made to measure phase, although in principle this could be done by comparing the phase of the signal from the detector with that from another detector and light source chopped by the rim of the sector disk.

This method is directly applicable to polychromatic OTF measurement. It suffers, however, from the low light intensity from the pinhole object used, and, because of the finite extent of the sector disk, no measurement can be made at zero spatial frequency.

Baker (1965) used a square-wave radial grating as an object. This was imaged by means of an auxiliary lens onto an object slit (see Fig. 15) which limited the region of the

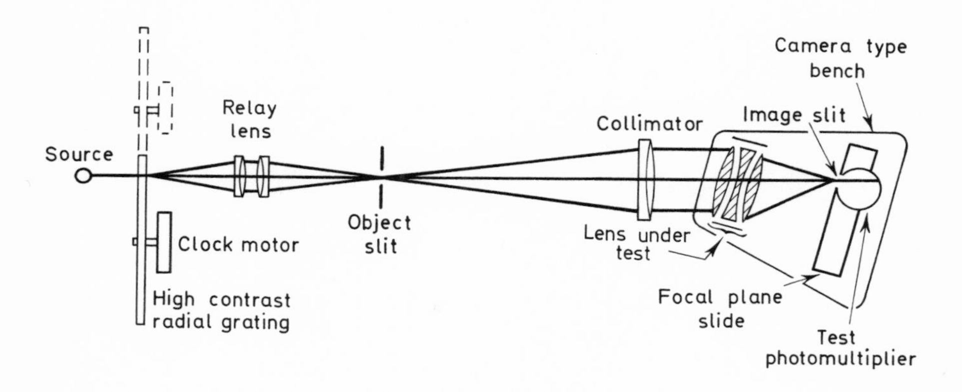

FIG. 15. Apparatus used by Baker.

grating seen by the lens under test. A further slit was set in the image plane of the test lens, at right angles to the object slit. The result of this effective superposition of two slits and the grating is shown in Fig. 16. If the grating is now made to rotate at constant angular velocity the effective scanning aperture is chopped at a constant temporal frequency. The effective fundamental spatial frequency of the grating depends on the angle θ between the direction of movement of the grating and the object slit, being zero when θ is zero and reaching a maximum when the edges of the grating are parallel to the object slit. This angle is made continuously variable by allowing the centre of the radial grating to be rotated about the optical axis of the system. The effective waveform of the grating is only square when

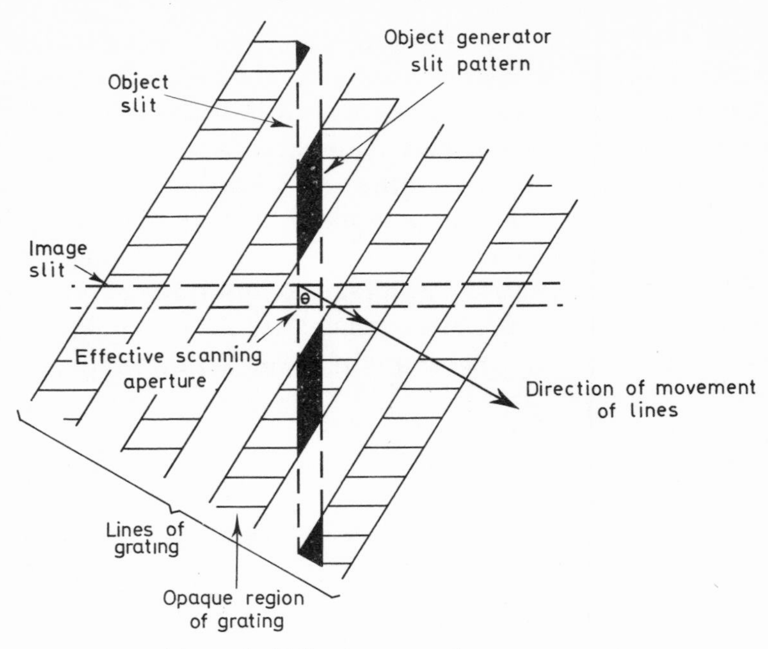

FIG. 16. Effective scanning aperture

its edges are running parallel to the object slit; otherwise it is a rhombic or triangular area grating. However, it may still be represented by a Fourier series, of which the fundamental term will be of constant frequency, and the detector must have a sufficiently narrow bandwidth to accept only the equivalent temporal frequency. Baker showed that, provided the slits are not wider than one quarter of the grating, the maximum change in amplitude of the fundamental with grating orientation is less than 0·1 per cent.

In order to measure phase and take account of source intensity fluctuations a two-channel system was used. The modified equipment up to the object slit is shown in Fig. 17, and on the test lens side of the object slit the system is as before. A zoom lens is used to focus the radial grating through a beam-splitter, the portion reflected being incident, via a pinhole, on a photomultiplier which provides phase and source intensity reference signals. The transmitted portion of the beam then passes through an adjustable phase-shifting plate, and the grating is finally focused by means of a relay lens onto the object slit.

The purpose of the phase-shifting plate is to ensure that the reference pinhole and the area common to the object and image slits both cover the same region of the grating, otherwise there will be an apparent linear variation of phase with spatial frequency. If the equipment is to be used with polychromatic light it is important to ensure that the phase-shifting plate is tilted, at the most, by a very small amount, or dispersion will cause the image to deteriorate by applying different lateral shifts to the image at different wavelengths. This calls for accurate alignment of the equipment.

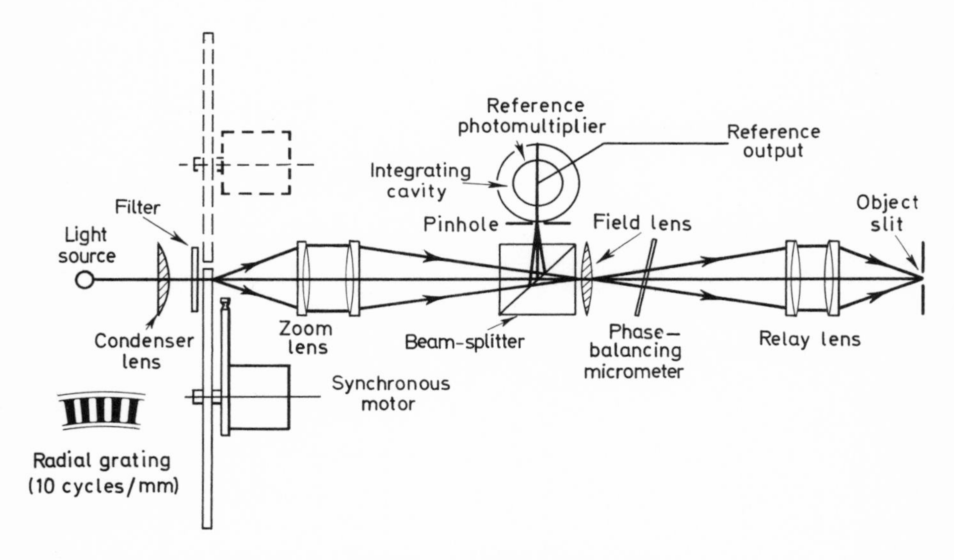

FIG. 17. Two-channel system of Baker.

The zoom lens allows one to choose the range of spatial frequencies over which the OTF is to be measured, by adjusting the magnification applied to the grating. For polychromatic use the zoom lens must be well corrected chromatically, as must the relay lens, although this correction need only be on axis.

This method has the advantage of continuous measurement down to zero spatial frequency so that the normalized OTF may be obtained and the accuracy of measurement of both modulus and phase of the OTF is high (approximately 1 per cent).

3.3. INTERFEROMETRIC METHODS

3.3.1. *Theory*

It is shown in Appendix 2 that the complex amplitude distribution in the image of a monochromatic point source is given by

$$K_\lambda(X', Y') = C \iint_{\text{pupil}} \exp \left\{ \frac{-2\pi i}{\lambda} \left(\frac{rX'}{R} + \frac{sY'}{R} + W_\lambda(r, s) \right) \right\} drds$$

where r, s are Cartesian coordinates in the exit pupil of the imaging system

R is the radius of the reference sphere, i.e. the distance of the Gaussian image point from the centre of the exit pupil

$W_\lambda(r, s)$ is the wavefront aberration and includes defocus.

This equation can be rewritten, ignoring the constant of proportionality, as follows:

$$K_\lambda(X', Y') = \iint_{\text{pupil}} k_\lambda(r, s) . \exp\left\{ \frac{-2\pi i}{\lambda} \left(\frac{rX'}{R} + \frac{sY'}{R} \right) \right\} drds \quad (3.11)$$

where

$$k_\lambda(r, s) = \exp\left\{ \frac{-2\pi i}{\lambda} . W_\lambda(r, s) \right\} \quad (3.12)$$

and is called the pupil function. It is further defined to be zero outside the region of the exit pupil, so that the limits of integration of the equation above may formally be set to $\pm\infty$. Then, converting to dimensionless coordinates by putting

$$u = \frac{r}{a} \qquad v = \frac{s}{a} \quad (3.13)$$

and

$$x' = \frac{aX'}{\lambda R} \qquad y' = \frac{aY'}{\lambda R} \quad (3.14)$$

where a is a dimension related to the pupil, equation (3.11) becomes

$$K_\lambda(x', y') = \iint_{-\infty}^{\infty} k_\lambda(u, v) . \exp\{-2\pi i(ux' + vy')\} dudv$$

This equation shows that k_λ is the Fourier transform of K_λ.

The image intensity distribution, the point spread function, is given by

$$P_\lambda(x', y') = |K_\lambda(x', y')|^2$$

and, by the autocorrelation theorem, taking the Fourier transform of both sides of this equation gives

$$p_\lambda(u_c, v_c) = \iint_{-\infty}^{\infty} k(u+u_c, v+v_c) . k^*(u, v)\, du dv \qquad (3.15)$$

where

$$u_c = \frac{1}{x_c'} = \frac{\lambda R}{aX_c'} = \frac{\lambda R}{a} . f_c$$

$$v_c = \frac{1}{y_c'} = \frac{\lambda R}{aY_c'} = \frac{\lambda R}{a} . g_c$$

and f_c and g_c are spatial frequencies of wavelength X_c' and Y_c' respectively.

Thus the OTF is the autocorrelation function of the pupil function. This relation holds only for one wavelength since the coordinate transformations of equations (3.14) are wavelength-dependent.

It should be noted here that the area of integration for equation (3.15) is the area common to the pupil centred on the point $u = 0$, $v = 0$, and its sheared counterpart centred on u_c, v_c, as shown in Fig. 18 for a circular pupil. Therefore,

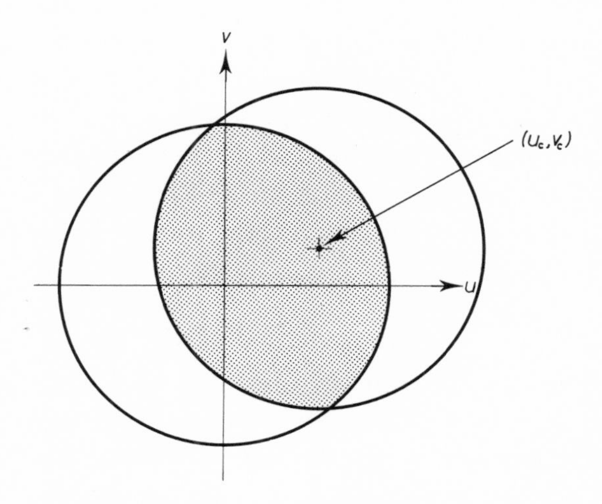

FIG. 18. Sheared exit pupil.

even for a wavefront with no aberration, the modulus of the OTF (the contrast transfer function) decreases because of the decreasing area as the spatial frequencies, given by u_c and v_c, increase. The rate of decrease of course depends upon the shape of the pupil, which may be quite intricate if vignetted.

Hopkins (1955a) showed that the autocorrelation function of the pupil function could be obtained using a wavefront-shearing interferometer, and that this method of measuring the OTF offered improved photometric efficiency over scanning methods.

Consider the exit pupil of an imaging system to be sheared by a suitable interferometer through transformed coordinate distances u_c and v_c, as shown in Fig. 18. Let the optical path difference between the two beams of the interferometer give rise to a phase difference δ between the two wavefronts. Then the total intensity of the interferogram is given by

$$I = \iint_{-\infty}^{\infty} |k_\lambda(u+u_c, v+v_c) + e^{-i\delta} . k_\lambda(u, v)|^2 \, dudv$$

$$= \iint_{-\infty}^{\infty} |k_\lambda(u+u_c, v+v_c)|^2 \, dudv + \iint_{-\infty}^{\infty} |k_\lambda(u, v)|^2 \, dudv$$

$$+ 2 \, \mathrm{Re}\{e^{-i\delta} \iint_{-\infty}^{\infty} k_\lambda(u+u_c, v+v_c) . k_\lambda^*(u, v) \, dudv$$

where Re stands for 'real part of'. The first two integrals of this equation are simply the intensities of the two beams and are constant for constant source intensity. Comparing the third term with equation (3.15) shows that it is proportional to the real part of the OTF. Using the notation of § 2.4 for the OTF, the third term may be written

$$2 \, \mathrm{Re}\{e^{-i\delta} . M_\lambda(u_c, v_c) . e^{i\varepsilon_\lambda(u_c, v_c)} = 2 \cos\{\varepsilon_\lambda(u_c, v_c) - \delta\} . M_\lambda(u_c, v_c)\}$$

If now the optical path in one arm of the interferometer is changed at a continuous rate of m wavelengths per second, the phase angle δ will change at a constant rate of $2\pi m$ radians per second, and the cosine term will cause the total intensity I to vary at a frequency of m cycles per second.

Thus, if the intensity is measured by means of a detector and narrow-bandwidth filter tuned to m Hz (to maximize the signal-to-noise ratio), the amplitude of the output and the phase angle relative to zero optical path difference will be proportional to the modulus and phase of the OTF respectively.

3.3.2. *Practical measurements*

Baker (1955) produced the first equipment to measure the OTF interferometrically, but it was extremely difficult to set up and highly susceptible to vibration. Kelsall (1959) improved on the method; his apparatus, shown schematically in Fig. 19, is basically a Michelson interferometer with corner-cube reflectors instead of plane mirrors. The plane

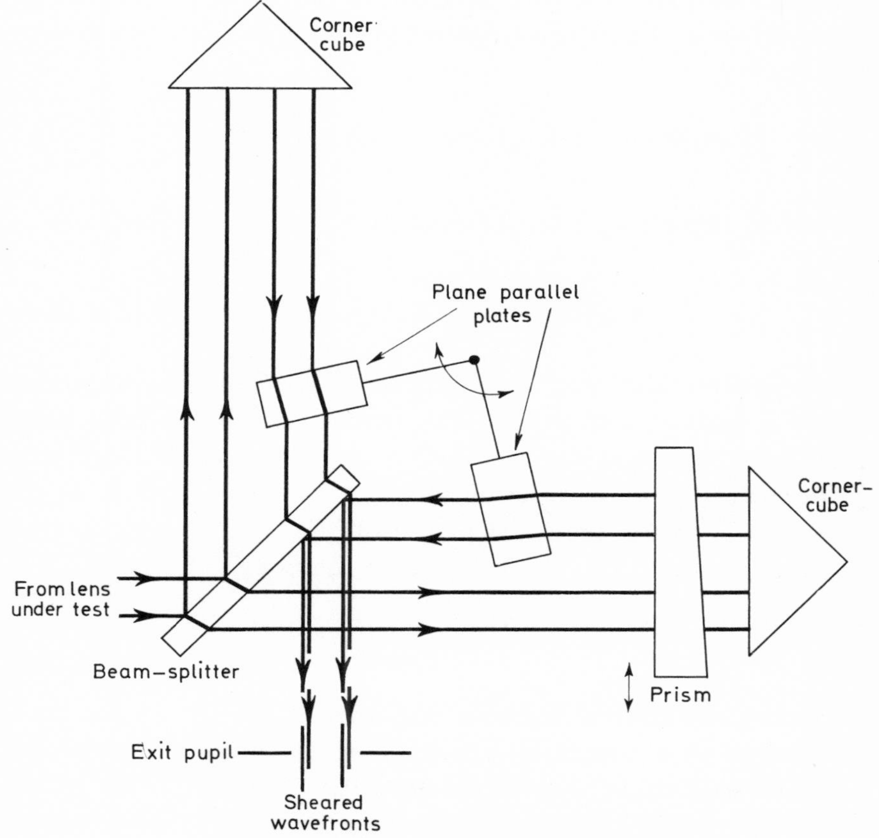

FIG. 19. Apparatus used by Kelsall.

parallel plates are slowly rotated to vary continuously the shear of the pupil, while the prism is made to oscillate back and forth comparatively rapidly, to provide the changing optical path.

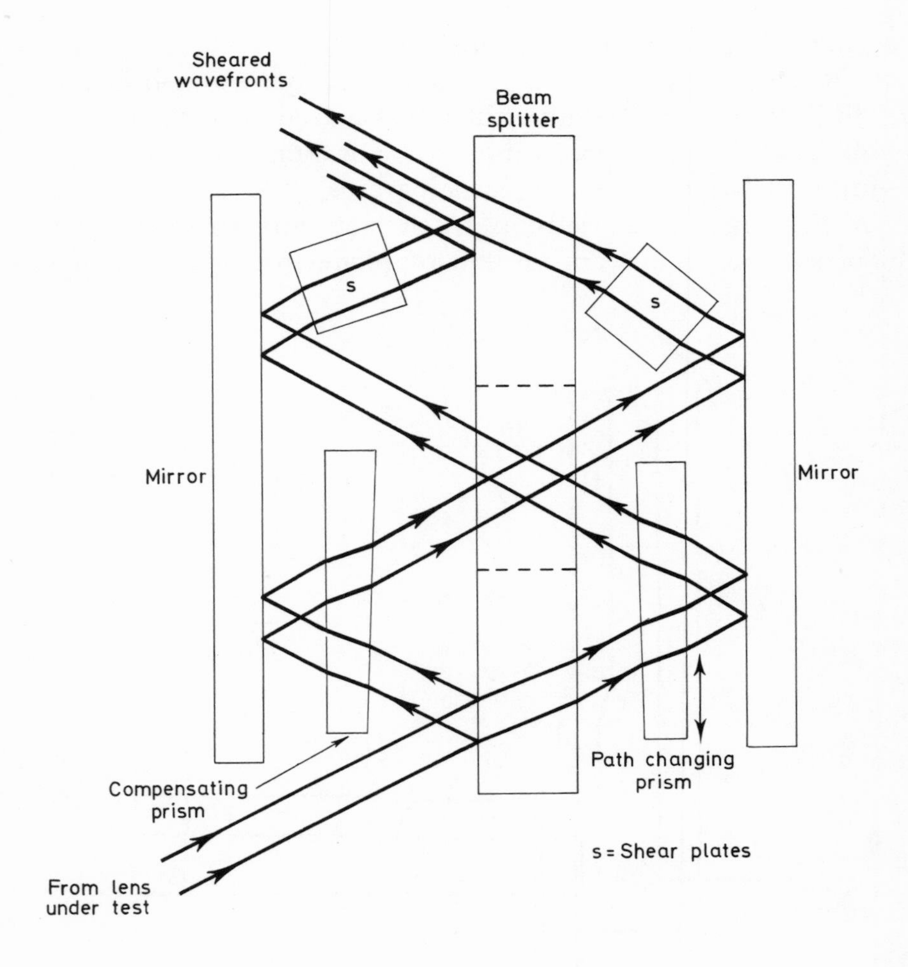

FIG. 20. Apparatus used by Montgomery.

Montgomery (1964) developed a more stable and easily adjusted shearing interferometer, shown schematically in Fig. 20. As before, the shear is applied by means of tilting plates and the optical path length is changed in one arm by means of an oscillating prism. Another fixed prism was included in the other arm to compensate for the gross path difference; this is necessary because of the finite coherence length of practical monochromatic sources.

It has been suggested by Jamieson (1968) that, although these interferometric methods cannot give a polychromatic OTF, if they were used with a polychromatic source they could be made to give the monochromatic OTF at various wavelengths by selection of the appropriate detector filter passband. For example, suppose the optical path in one arm of the interferometer were being changed by a wedge of glass of refractive index n and the thickness of this wedge were being changed at a rate of d microns per second. Then for a detector tuned to ν Hz, the effective wavelength in use would be λ, where

$$\frac{n_\lambda - 1}{\lambda} = \frac{\nu}{d}$$

(n_λ being the refractive index at wavelength λ), and because of the physical characteristics of glasses this relation may only be satisfied for one wavelength.

For a constant shear u_c ($v_c = 0$), the true spatial frequency at which the OTF is measured would then be given by

$$f_c = \frac{a.u_c}{\lambda R} = \frac{a.u_c}{R} \cdot \frac{\nu}{d(n_\lambda - 1)}$$

This system would seem quite practicable, but it has two undesirable features. Firstly, from consideration of the required temporal coherence of the effective source, it would be necessary to have a sufficiently narrow-bandwidth detector. If the bandwidth were too large, giving an effective source having a large spectral width, the coherence length would be too short for sufficient cycles of the detector output to be obtained before the contrast of the interference pattern was reduced. Secondly, for a given detector bandwidth, the increased total intensity falling on the detector in the polychromatic case will cause the signal-to-noise ratio to be less than if a monochromatic source were used.

3.4. SUMMARY

The scanning methods can in principle all be used to measure the polychromatic OTF. However, to do so, it is necessary to ensure that any auxiliary lenses used are chromatically corrected, which adds to the complexity and cost of the equipment. This problem is worst for the more sophisticated methods of measurement, which are capable of higher monochromatic accuracy and are in general more flexible, since they tend to employ more auxiliary lenses. It is also, of course, necessary to have available sources and detectors which have the required spectral characteristics, and this can be a problem.

The interferometric methods can only give measurements of the monochromatic OTF.

4

Methods of Computation

4.1. INTRODUCTION

To enable the performance of a lens to be predicted, and to check any measurements made, it is necessary to calculate the OTF from the optical design data.

The most accurate calculations are based on diffraction theory, but these tend to be very intricate, taking a long time to complete even with high-speed digital computers. It has been established that for lenses having a wavefront aberration of several wavelengths or more, much simpler calculations based on geometrical optics may be used to give a sufficiently good approximation to the OTF. For wavefront aberrations below a few wavelengths a useful approximation is obtained from the product of the OTF for a perfect lens and the OTF as computed by geometrical optics methods.

The diffraction theory calculations can only give monochromatic OTFs, so that it is necessary to perform the calculation at several wavelengths and then take the sum of these OTFs, suitably weighted for source, detector and transmission spectral characteristics, as indicated in § 2.3.2. The geometrical optics method can be used to give the polychromatic OTF directly.

4.2. DIFFRACTION THEORY METHOD

The monochromatic OTF has previously been shown to be the autocorrelation function of the pupil function (equation (3.15)). By a translation of axes this equation may be written symmetrically, as follows:

$$p_\lambda(u_c, v_c) = \iint\limits_{-\infty}^{\infty} k_\lambda\left(u+\frac{u_c}{2}, v+\frac{v_c}{2}\right).k_\lambda^*\left(u-\frac{u_c}{2}, v-\frac{v_c}{2}\right)\mathrm{d}u\mathrm{d}v$$

or, substituting for k_λ from equation (3.12) ,

$$p_\lambda(u_c, v_c) = \iint\limits_{-\infty}^{\infty} \exp\left\{-\frac{2\pi i}{\lambda}\left[W_\lambda\left(u+\frac{u_c}{2}, v+\frac{v_c}{2}\right) - W_\lambda\left(u-\frac{u_c}{2}, v-\frac{v_c}{2}\right)\right]\right\}\mathrm{d}u\mathrm{d}v \qquad (4.1)$$

The region of integration is the area common to the sheared pupils, as shown in Fig. 21, for a circular pupil, since k is zero by definition outside the pupil.

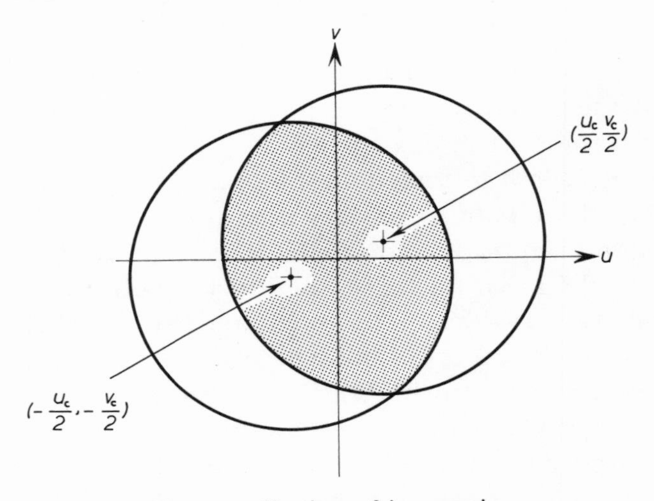

FIG. 21. Region of integration.

Several investigations of idealized optical systems with certain amounts of aberration have been made by substituting suitably for W_λ in equation (4.1) and then integrating numerically. Hopkins (1955b) dealt with the OTF of a defocused aberration-free lens, De (1955) examined astigmatism, and Goodbody (1958, 1960) considered primary and secondary spherical aberration and primary and secondary coma. In order to deal with real lenses, whose wavefront aberrations are in general combinations of various primary, secondary and higher order terms, the wavefront aberration may be obtained either by measurement (using, for example, a Twyman-Green interferometer) if the lens has been manufactured, or by computation from the design data.

Marchand and Phillips (1963) calculated the wavefront aberration coefficients as follows. First, the vignetted aperture of the lens was computed and a number of rays distributed evenly over this aperture were traced from the object point to the exit pupil plane. In practice, it was found that 50 rays enabled sufficient accuracy to be achieved. The optical path along each of these rays was calculated and the departure from the lengths which would be expected for a spherical wavefront centred on the ideal image point were found. (The ideal image point was taken as the point at which the principal ray intersected the Gaussian image plane, since this removes large amounts of phase shift from the OTF if distortion is present). These differences were then fitted by a least-squares method to an aberration polynomial, giving W_λ.

The numerical integration of equation (4.1) was done by a method suggested by Hopkins (1957). The vignetted aperture was divided into squares each of side $2e$, the aberration function W_λ expanded as far as the first three terms of its Taylor series for a point at the centre of one of these squares, and the integration carried out over the area of that square. This was repeated for all the other squares, and the results were summed. For $e = 0{\cdot}05$ this method gives an accuracy of 5 per cent or better.

Bray (1965) points out that if the calculation above is used to obtain the polychromatic OTF from several monochromatic calculations, it is necessary to take into account the possible lateral shift of the ideal image points, as defined above, for various wavelengths caused by transverse chromatic aberration.

Barakat (1962) recommended a different approach for rotationally symmetric aberrations, which is capable of much greater accuracy, especially at high spatial frequencies or with large aberrations. The disadvantage of the least-squares fitting is that the accuracy of fit over the whole of the wavefront is not uniform. To overcome this Barakat fitted Chebyschev polynomials to the path differences calculated as above. These, unlike least squares, may be made to give any required accuracy provided that enough terms are taken. As a further improvement the numerical integration of equation (4.1) was done by Gaussian quadrature, which is economic in computation time. The overall accuracy for this method was shown to be 0·01 per cent.

4.3. GEOMETRICAL OPTICS METHOD

The domain of geometrical optics is reached when the wavelength of light is assumed to tend to zero. The exponential term in equation (4.1) may be written in terms of spatial frequencies as follows:

$$-\frac{2\pi i}{\lambda}\left[W_\lambda\left(u+\frac{\lambda R f_c}{2a}, v+\frac{\lambda R g_c}{2a}\right) - W_\lambda\left(u-\frac{\lambda R f_c}{2a}, v-\frac{\lambda R g_c}{2a}\right)\right]$$

This may be expanded by means of Taylor's series to give

$$-\frac{2\pi i}{\lambda}\left[2\left(\frac{\lambda R f_c}{2a}\cdot\frac{\partial}{\partial u}+\frac{\lambda R g_c}{2a}\cdot\frac{\partial}{\partial v}\right)W_\lambda(u, v) + \frac{2}{6}\left(\frac{\lambda R f_c}{2a}\cdot\frac{\partial}{\partial u}+\frac{\lambda R g_c}{2a}\cdot\frac{\partial}{\partial v}\right)^3 W_\lambda(u, v)+\ldots\right]$$

and as λ tends to zero this tends to

$$-2\pi i\left(\frac{R f_c}{a}\cdot\frac{\partial}{\partial u}+\frac{R g_c}{a}\cdot\frac{\partial}{\partial v}\right)W_\lambda(u, v)$$

Thus, equation (4.1), which now gives the geometric OTF, becomes

$$p_{\lambda,\,\mathrm{geom}}(f_c, g_c) = \iint_{-\infty}^{\infty} \exp\left\{-2\pi i\left(\frac{R f_c}{a}\cdot\frac{\partial}{\partial u}+\frac{R g_c}{a}\cdot\frac{\partial}{\partial v}\right)W_\lambda(u, v)\right\} \mathrm{d}u\,\mathrm{d}v \qquad (4.2)$$

where $\frac{R}{a}\cdot\frac{\partial}{\partial u}W_\lambda(u, v)$ and $\frac{R}{a}\cdot\frac{\partial}{\partial v}W_\lambda(u, v)$ are the transverse ray aberrations at image distance R (see Hopkins (1950)), and may be computed by ray tracing. Also, as the wavelength tends to zero, the shear of the pupil becomes zero. Thus the region of integration becomes the area of the pupil for all spatial frequencies, and equation (4.2) may be evaluated.

It should be noted that although it has been assumed that the wavelength tends to zero, equation (4.2) only gives the

OTF at one wavelength since W_λ is still a function of wavelength. The calculation of the transverse ray aberrations must therefore be repeated for various wavelengths, and the values obtained suitably weighted to take account of the spectral characteristics of the source, detector and system transmission, in order to obtain the geometrical optics approximation to the polychromatic OTF.

This is equivalent to obtaining spot diagrams (first proposed by Herzberger (1947)) for various wavelengths, superimposing them after suitable weighting to give an approximation to the polychromatic point spread function, and then performing a Fourier transform to obtain the OTF.

As a further approximation, Hopkins has defined an 'average-circular' geometric transfer function which is obtained by taking the Fourier transform of a rotationally averaged spot diagram. This gives an OTF which is in effect an average over all azimuths in the image plane.

4.4. COMPARISON OF METHODS

Several comparisons have been made between results obtained using the geometrical optics approximation and diffraction theory based calculations. Hopkins (1955b) discussed defocus of a perfect system, De (1955) dealt with astigmatism, Bromilow (1958) with defocus and primary and secondary spherical aberration, and Marathay (1959) with coma. Broadly their conclusions were that for an accuracy in the calculated OTF of about 5 per cent of the zero spatial frequency value, over the range of frequencies met with in most photographic lenses, the geometrical optics approximation is sufficient provided that the wavefront aberration exceeds 1–2 wavelengths.

The difference between the diffraction theory based and geometrical optics calculations of OTF arise from the different areas of integration used and the neglected higher order terms of the Taylor series expansion of the wavefront aberration. For small aberrations (W less than about two wavelengths) the former predominates, and a useful approximation is obtained by multiplying the OTF obtained from the geometrical optics calculation by that of a perfect system of the same aperture. See for example Miyamoto (1961).

Miyamoto also discussed the case of large aberrations (W greater than two wavelengths) and concluded that, provided high spatial frequencies are not of interest, the geometrical optics approximation gives good results.

To give an idea of the saving in time achieved by the geometrical optics approach Bray (1965) used Hopkins approximation to compute polychromatic OTFs for an f/3·5, 24 inch aerial lens, achromatized over the range 0·6–0·7 micron and known to be limited by secondary spectrum. The computer time taken was 6½ minutes, which compares very favourably with the three hours taken to do the same calculations on the same computer by the diffraction based method of Marchand and Phillips discussed previously. The results of the two methods agreed to within approximately 6 per cent of the zero spatial frequency value up to a spatial frequency of 60 lines per millimetre.

5

Examples of Polychromatic Optical Transfer Functions

The object of this chapter is to give the reader some idea of the ways in which polychromatic OTFs may be calculated and used.

As a first example take the relatively simple case of a large refracting telescope objective which has no aberration on axis except for the residual variation of focal length with wavelength, known as the secondary spectrum. The Newall telescope at Cambridge fits this category, and its secondary

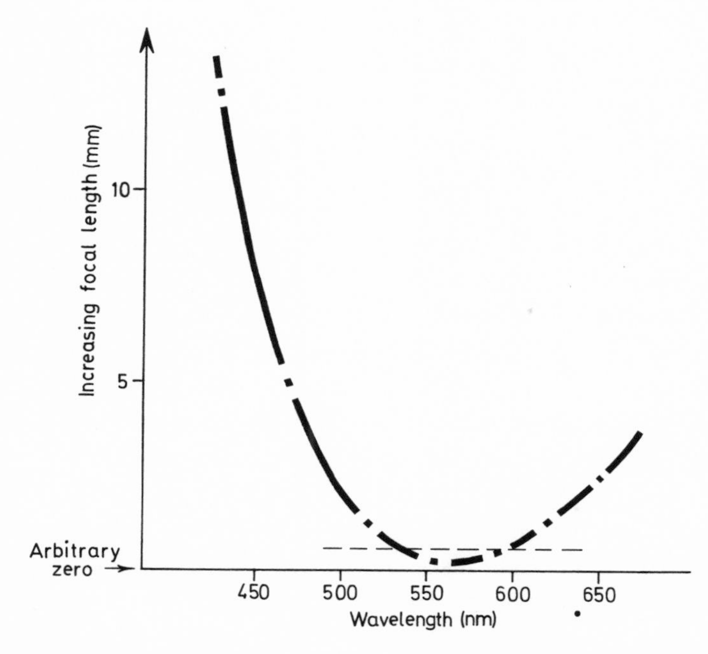

FIG. 22. Secondary spectrum curve of the Newall telescope.

spectrum curve is shown in Fig. 22. This objective has an aperture of 24·9 inches and a focal length of 29 feet. It was studied by Linfoot and Wolf (1952) to determine the best focal plane for use visually when observing stars of colour temperature 6000° K. Using two separate criteria, one of maximum visual brightness at the centre of the diffraction image and the other of maximum visual illumination within a circle of radius equal to that of the first dark ring of the Airy pattern at 566 nanometres, they concluded that this plane was 0·26 mm outside the minimum focal distance of Fig. 22.

To decide whether this focal position is also optimum for planetary detail on axis, a series of OTF curves through focus are required. First, the decision must be made whether to use a diffraction based OTF calculation or the approximation of geometrical optics. In this case this is easily decided because the lens by definition has no wavefront aberration on axis, and therefore the former method must be employed. Knowing the optical design data of the lens (the curvatures, glasses, separations and diameters) it would be possible to use a method such as that of Marchand and Phillips (Chapter 4) to evaluate the OTF, but this is not necessary here. Levi and Austing (1968) have produced tables of accurately calculated values of the OTF for a perfect lens with a circular pupil and varying amounts of defocus. Now for a given focal plane the amount of defocus at a particular wavelength may be obtained from the secondary spectrum curve in Fig. 22, and the contribution to the OTF at this wavelength may be found from the tables. This figure has then to be multiplied by the weighting factor $R_\lambda . S_\lambda . T_\lambda$ to give the integrand of equation (2.9).

R_λ is the spectral sensitivity of the receiver, which in this case is the eye, and values for this may be found from tables, e.g. in Levi (1969). S_λ is the spectral distribution of energy from the source, in this case a black body at 6000° K, and figures for this may also be found from tables. T_λ is the spectral transmission of the system and in this case is taken as unity. The product $R_\lambda . S_\lambda . T_\lambda$ is shown plotted against wavelength in Fig. 23.

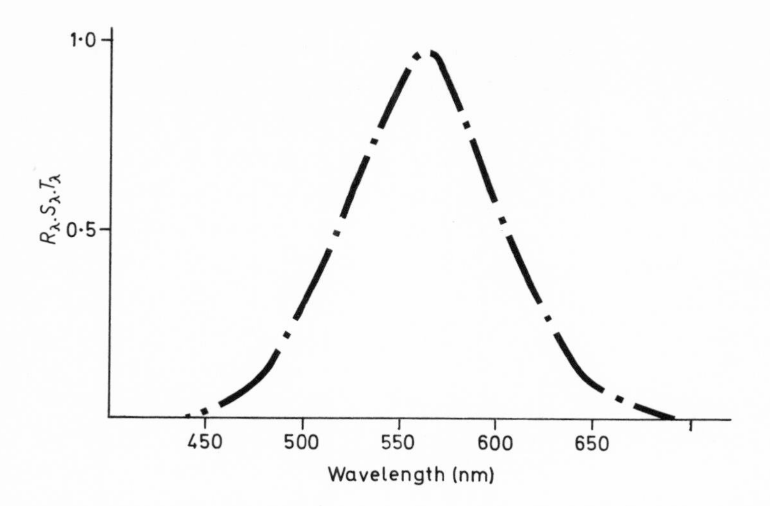

FIG. 23. Normalized product of source and detector spectral characteristics.

Curves of weighted monochromatic OTF values at various spatial frequencies and various focal plane positions may then be calculated; two such curves are shown in Fig. 24 for two

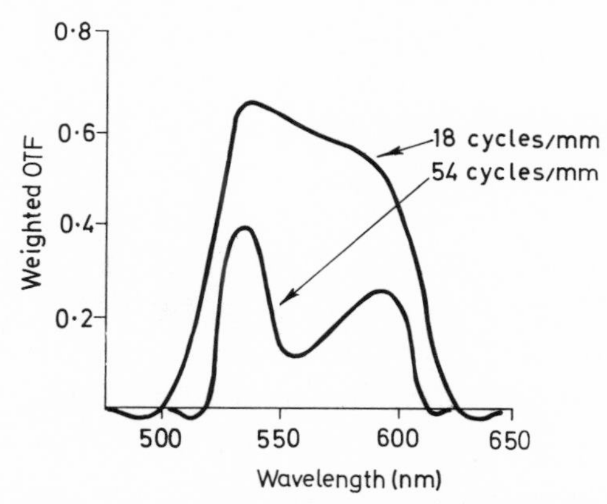

FIG. 24. Variation of weighted OTF with wavelength at constant spatial frequency.

spatial frequencies. The area under each curve represents the integral of equation (2.9), which is the polychromatic OTF at the spatial frequency specified. The shapes of these curves are interesting, especially that at the higher spatial frequency. The focal plane to which these curves refer is shown as a broken line on the secondary spectrum curve in Fig. 22, and it can be seen that there is perfect focus at two wavelengths, giving rise to the peaks in the weighted OTF curves.

A series of such curves was drawn for several spatial frequencies at the best focal position found by Linfoot and Wolf and also at focal positions 0·3 mm on either side, and the polychromatic OTFs were calculated and normalized. The results are shown in Fig. 25. It can be seen that the focal

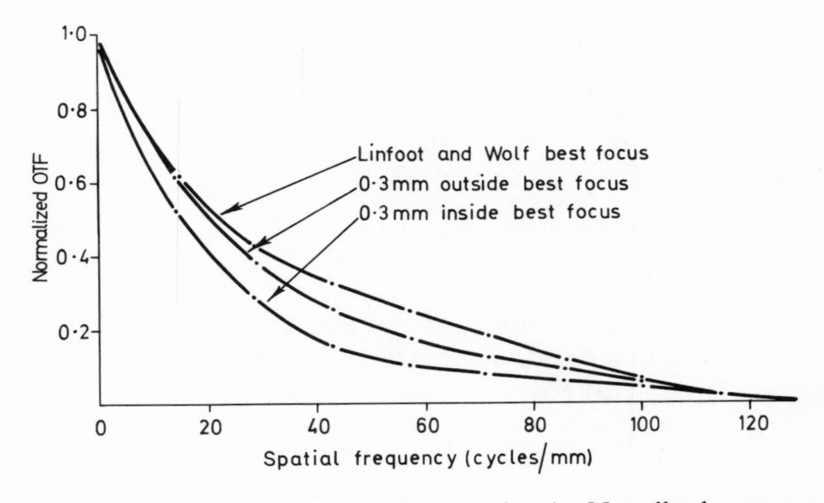

FIG. 25. Polychromatic OTF curves for the Newall telescope.

plane selected by Linfoot and Wolf is the best of the three out to the highest spatial frequencies. To give some idea of the spatial frequencies involved, markings on the planet Mars having a spatial period one fortieth of the planet's diameter would correspond to a spatial frequency of 50 c/mm in the image at the planet's mean opposition distance from Earth. At this spatial frequency the selected focal plane is significantly superior to the other two.

As a more complex example of a polychromatic OTF calculation take the lens whose construction and design data are shown in Fig. 26. This lens was described in a paper by Hopkins and Unvala (1966) and is one of several derived by various computer optimization programs from a single original design. The lens has a focal length of 24 inches, a relative aperture of f/6·0 and is designed to cover a 9 × 9 inch

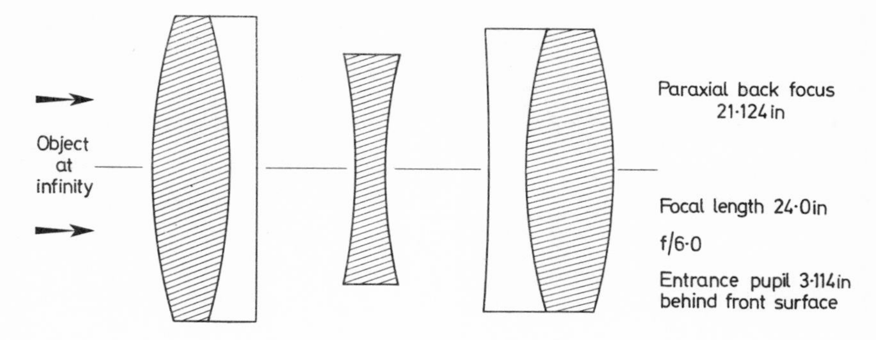

Radius	*Thickness*	*Clear diameter*	*Refractive index*	*V-value*	*Material*
+7·977		4·26			
	1·020		1·62041	60·3	SK16
−8·765					
	0·400		1·51895	57·4	K4
−751·5					
	1·323				Air
−8·172					
	0·400		1·57501	41·5	LF7
+7·241		3·26			
	1·371				Air
−29·13					
	0·400		1·51895	57·4	K4
+7·085					
	1·160		1·62041	60·3	SK16
−7·225		3·88			

FIG. 26. Construction and design data of lens (after Hopkins and Unvala, 1966).

square format. It is colour corrected over a wavelength range of 480 to 650 nanometers.

Let us suppose that this lens is to be used in an aerial survey camera and it is required to know its performance in terms of the OTF. To do this it is first necessary to stipulate its exact working conditions, in terms of source, detector and system spectral transmission. For the source let us assume that it is the Sun and that this may be approximated by a black-body radiator at a temperature of 5000° K. Levi (1969) gives a table of the normalized output from such a

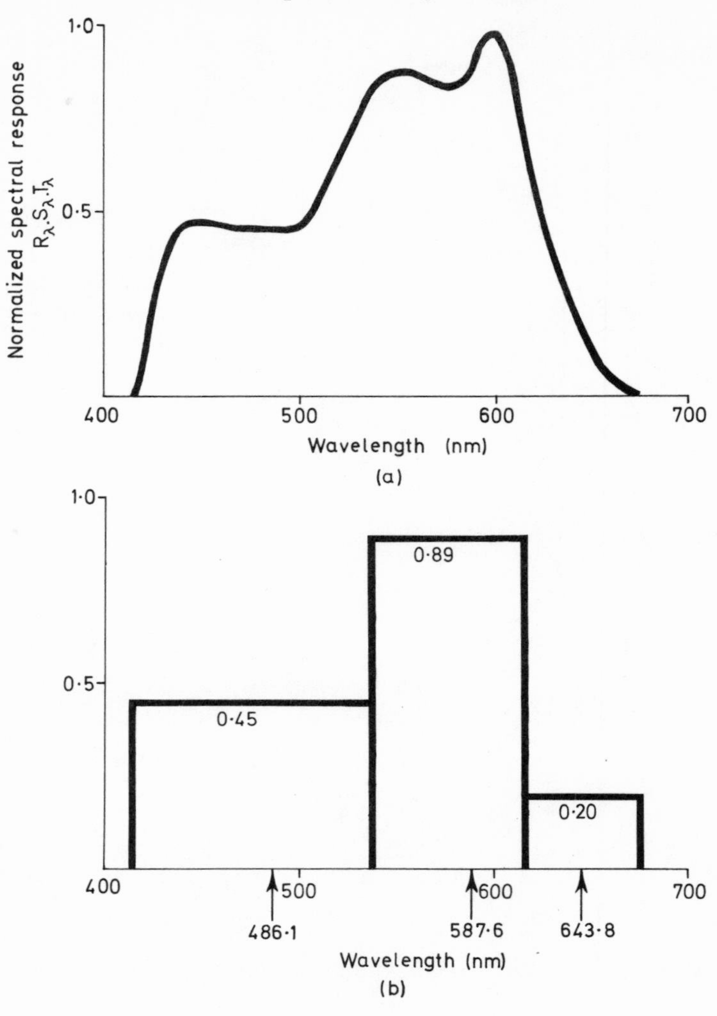

FIG. 27. (*a*) Normalized spectral response curve. (*b*) Equivalent rectangular response curve.

body, and in the calculations that follow his values will be used. The detector will be taken as panchromatic film, and values of the spectral sensitivity of such a film are again taken from Levi. In order to reduce the effects of ultra-violet radiation scattered from the atmosphere when the lens is used at high altitude, it is necessary to use a filter which absorbs ultra-violet and some blue but passes the longer wavelengths of the visible spectrum. A suitable filter is Kodak Wratten

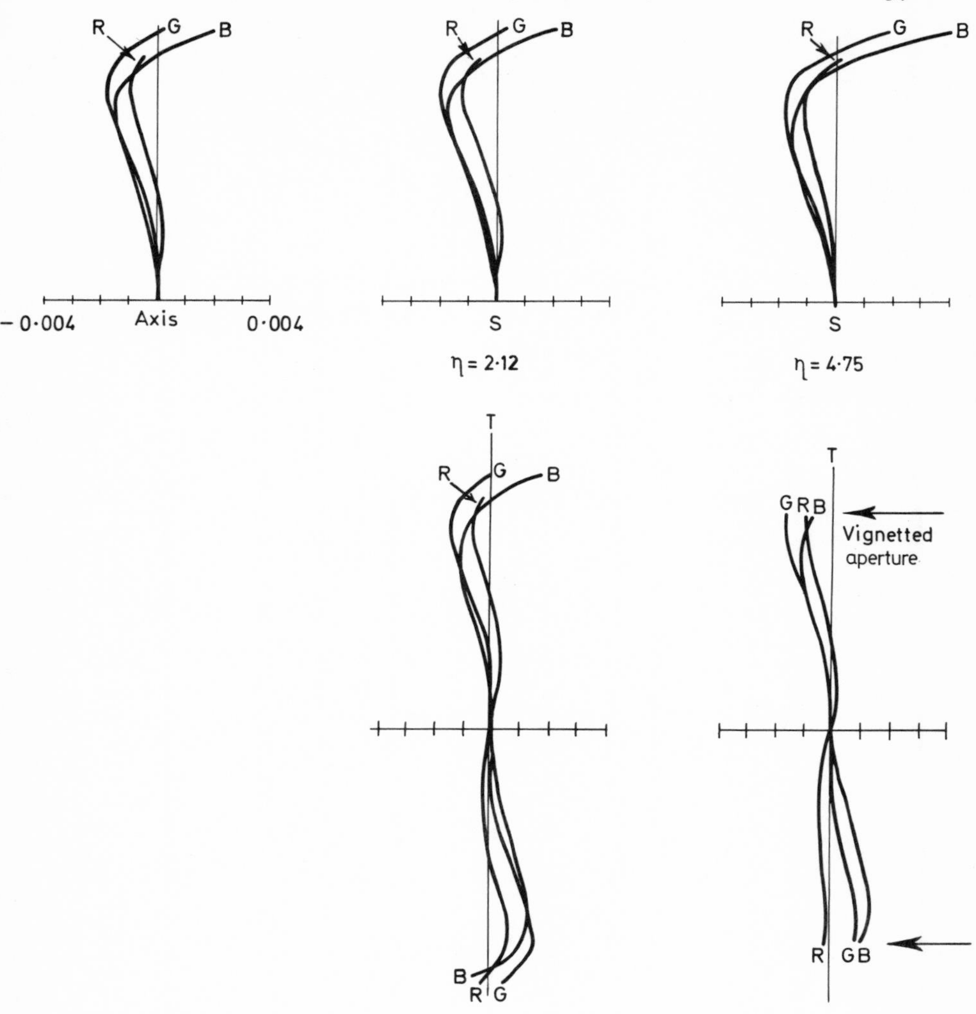

FIG. 28. Transverse ray aberrations (inches). Wavelengths: G 587·6 nm; R 643·8 nm; B 486·1 nm.

No. 2E, and the combination of figures for its transmission at various wavelengths (obtained from Kodak data sheet FT9, issue B) together with the transmissions of the glasses in the lens (obtained from the axial thicknesses of the lens elements and the figures of transmission given in the Schott data sheets for the glasses concerned) gives the system spectral transmission. The product of the source, receiver and transmission characteristics is shown graphically in Fig. 27(*a*).

The next task is to decide between a diffraction theory based calculation of the OTF and a geometrical optics approximation. To do this it is necessary to have some idea of the aberrations of the lens, and this is conveniently and most usually obtained by calculating transverse ray aberrations. (A survey of the methods available for calculation and presentation of lens aberration data is given by Palmer (1971)). A set of transverse ray aberrations in three wavelengths, on axis and at two field positions for sagittal (S) and tangential (T) fans was computed for the lens in question, and is shown in Fig. 28. The field radii chosen were 2·12 and 4·75 inches, and the relation between these and the total image format is shown in Fig. 29.

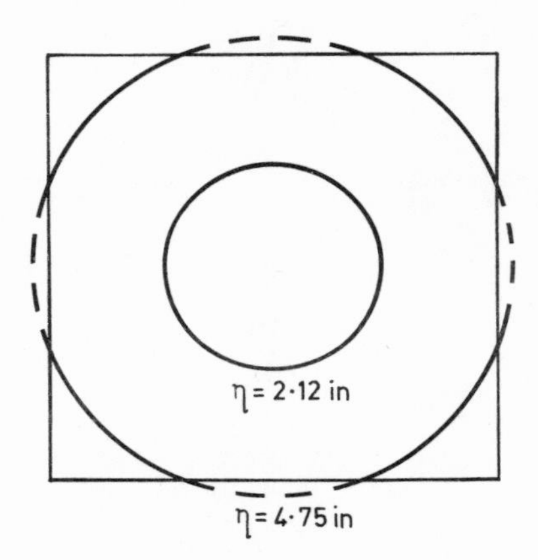

FIG. 29. The relation between the two field radii selected and the total format.

Considering first the axial aberrations in Fig. 28 it can be seen that the radius of the aberration patch would be approximately 0·002 inches, which is about twelve times the radius of the first dark ring of the Airy pattern in green light for a perfect system of the same focal length and aperture. However this patch size may be reduced by defocusing. If the film plane is brought closer to the lens by 0·012 inch then a term which is linearly dependent on aperture and has a maximum value of 0·001 inch at the edge of the aperture is added to the transverse aberration. This has the effect of reducing the radius of the aberration patch to a minimum of about 0·001 inch, which is six times the radius of the Airy disk. This is large enough to indicate that the approximation of geometrical optics should be sufficient. The fact that the aberrations at the two field positions selected are very similar to those on axis shows that this conclusion applies equally over the field, and also indicates that since the aberrations are varying only slowly with field, the two field positions chosen should be enough to give a reasonably accurate impression of the performance of the lens over most of the working field.

We may now proceed to compute monochromatic OTFs at various wavelengths within the range of interest at required spatial frequencies, and draw curves of OTF against wavelength for the axis and the two field positions chosen. Then the product of these curves with that of Fig. 27(*a*) gives the integrand of equation (2.9), and the area under the combined curve gives the polychromatic OTF. This was done using a geometrical optics OTF computer program, at seven wavelengths, four planes of focus and spatial frequencies of 200 and 600 c/inch (which correspond to spatial frequencies on the ground of 1 cycle per 60 and 20 inches respectively if the lens is used at a height of 24 000 ft.). The results are shown in Figs 30(*a*) and 31(*a*).

At the two field positions, OTFs were calculated for two grating orientations, radial (or sagittal), S, and tangential, T. Since the aberration in the latter direction is asymmetric, phase effects appear and it is necessary to perform twice as many calculations to obtain both the polychromatic MTF and phase. In all 56 graphs of OTF against wavelength were necessary.

A much quicker method, which is capable of reasonable accuracy provided that the OTF does not change greatly with wavelength is as follows. Taking, say, three wavelengths spaced out through the spectral range of interest, divide the intervals between adjacent wavelengths in half. At these inter-

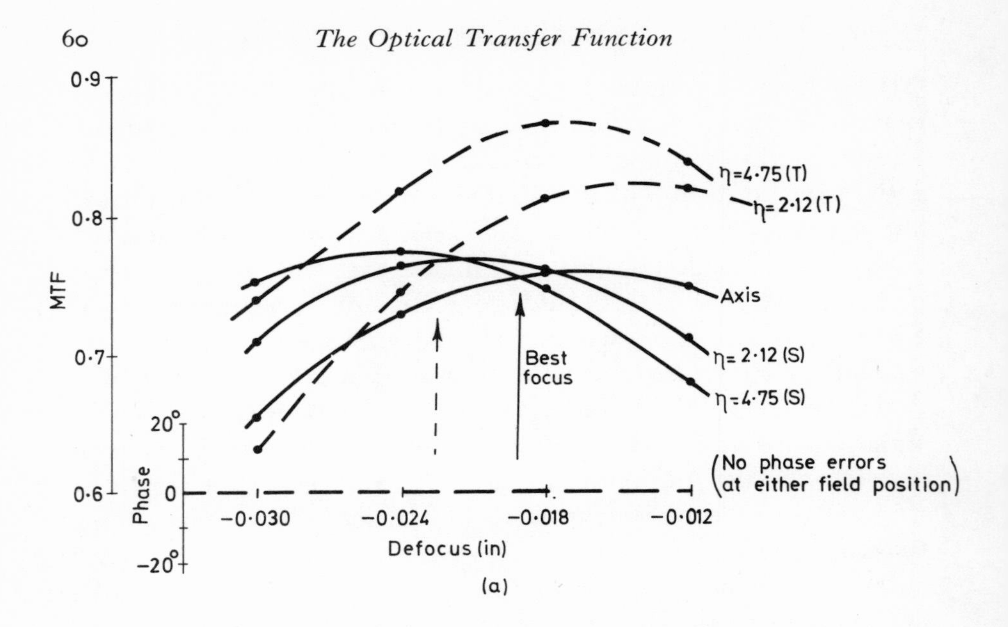

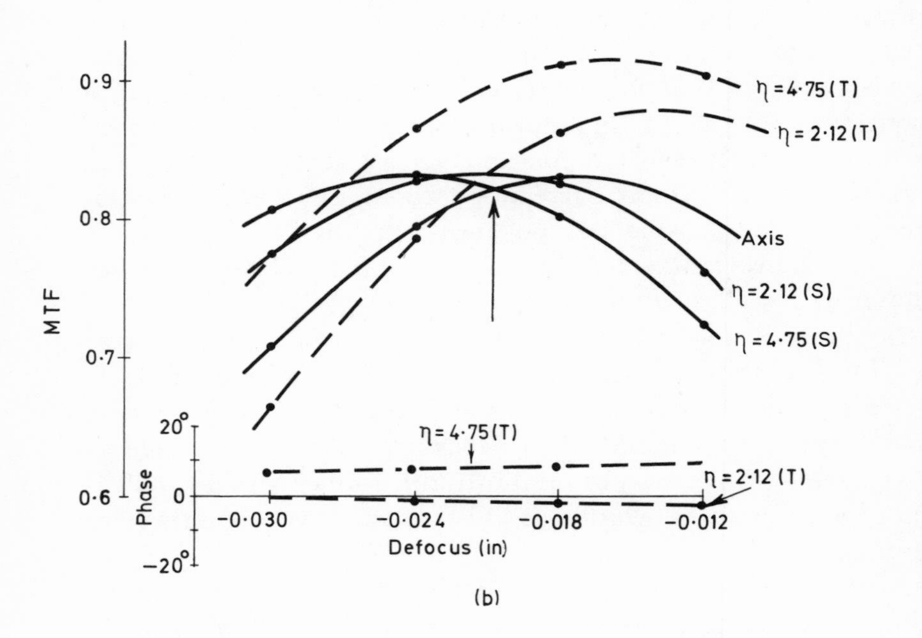

FIG. 30. Graphs of OTF through focus at 200 c/inch spatial frequency. (*a*) Accurate polychromatic OTF. (*b*) Three-wavelength approximate polychromatic OTF.

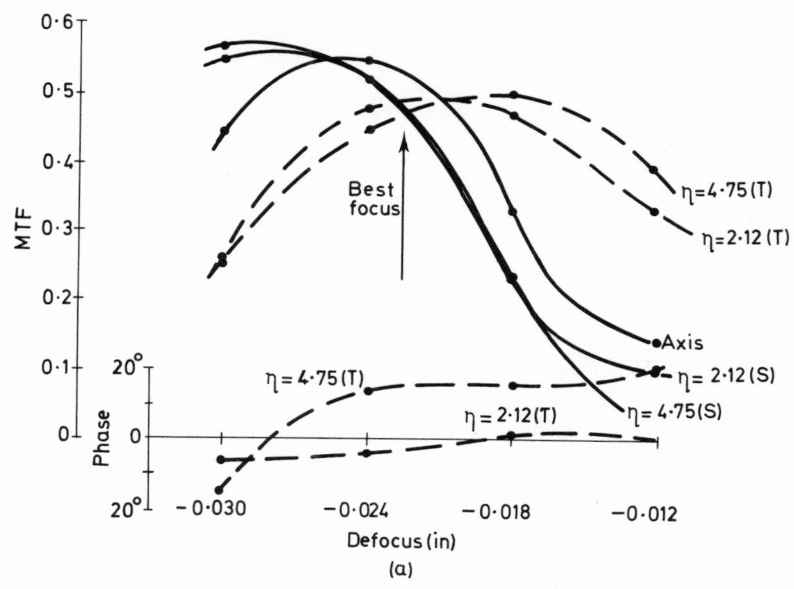

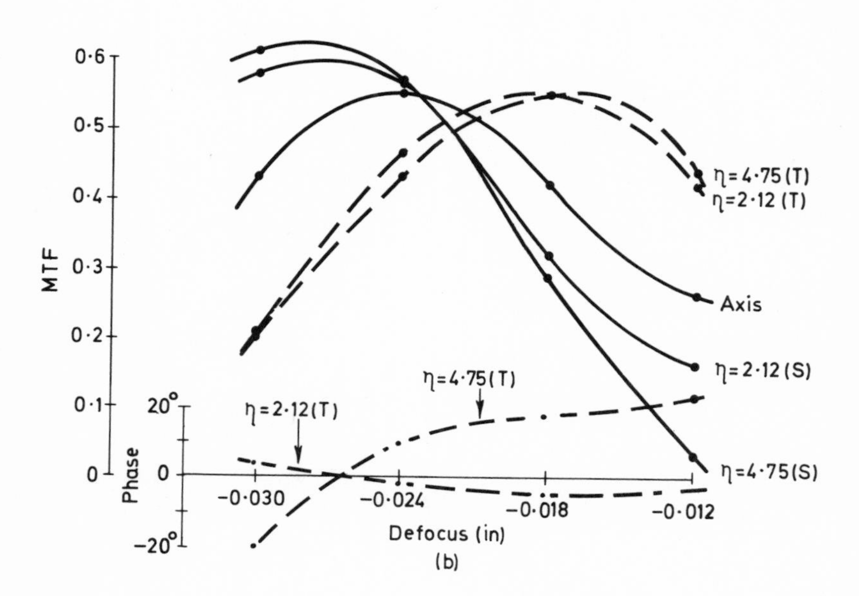

FIG. 31. Graphs of OTF through focus at 600 c/inch spatial frequency. (*a*) Accurate polychromatic OTF. (*b*) Three-wavelength approximate polychromatic OTF.

mediate wavelengths on the spectral response curve, e.g. Fig. 27(*a*), draw vertical lines to divide the curve into three areas, and on the base of each area construct a rectangle having the same area as the area under the curve. The heights of these rectangles are then the weighting factors to be used on the OTFs obtained at the three original wavelengths, and the sum of these weighted OTFs then approximates the polychromatic OTF.

This was done for the lens being considered, and the equivalent 'rectangular' spectral response curve obtained is shown in Fig. 27(*b*). It happens in this case that the weighting factors so obtained coincide almost exactly with the values of the normalized spectral response curve of Fig. 27(*a*) at the wavelengths selected, but in general this would not be the case. Using these weighting factors on the OTFs obtained at the three specified wavelengths, sets of polychromatic OTF curves through focus were drawn as before; they are shown in Figs. 30(*b*) and 31(*b*). It can be seen that the results obtained are generally similar to those obtained by the more accurate method except for an apparent increase in level at 200 c/inch.

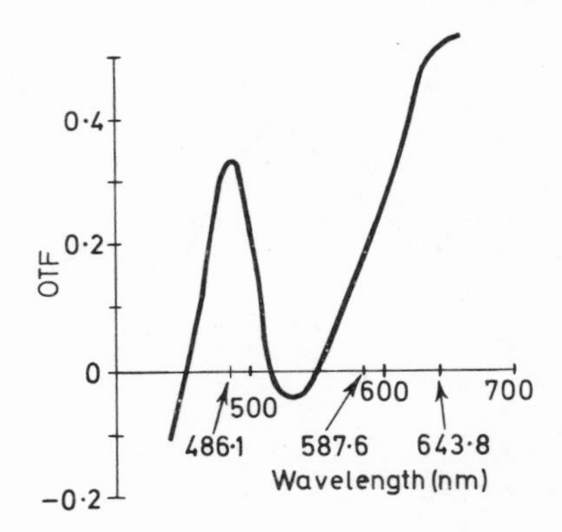

FIG. 32. Graph of variation of OTF with wavelength. On axis defocus −0·012 inch, spatial frequency is 600 c/inch.

One point where there is a large discrepancy is at 600 c/inch, on axis at −0·012 inch defocus. The accurate method gives an MTF value of 0·14 and the three-wavelength approximation gives a value of 0·27. On investigating the plot of OTF against wavelength for this position (see Fig. 32) it is found that it is by no means constant with wavelength, and the peak in the blue occurs almost exactly at the wavelength chosen, which causes the disparity.

To return to the curves of Figs. 30(*a*) and 31(*a*) obtained by the more accurate method, it can be seen that the best focal plane at 200 c/inch is obtained by defocusing by approximately −0·019 inch, and at 600 c/inch it is at −0·023 inch. Any departure from these positions may improve some parts of the field, but only at the expense of others. Since only one focal plane may be used it is necessary to strike some compromise between these two results. It is apparent from the slopes of the curves that the higher spatial frequency deteriorates more rapidly for a given displacement of the focal plane from its optimum position. It is therefore the position of best focus at the higher spatial frequency which is chosen as the overall best focal plane, and this is shown as the broken-line arrow in Fig. 30(*a*).

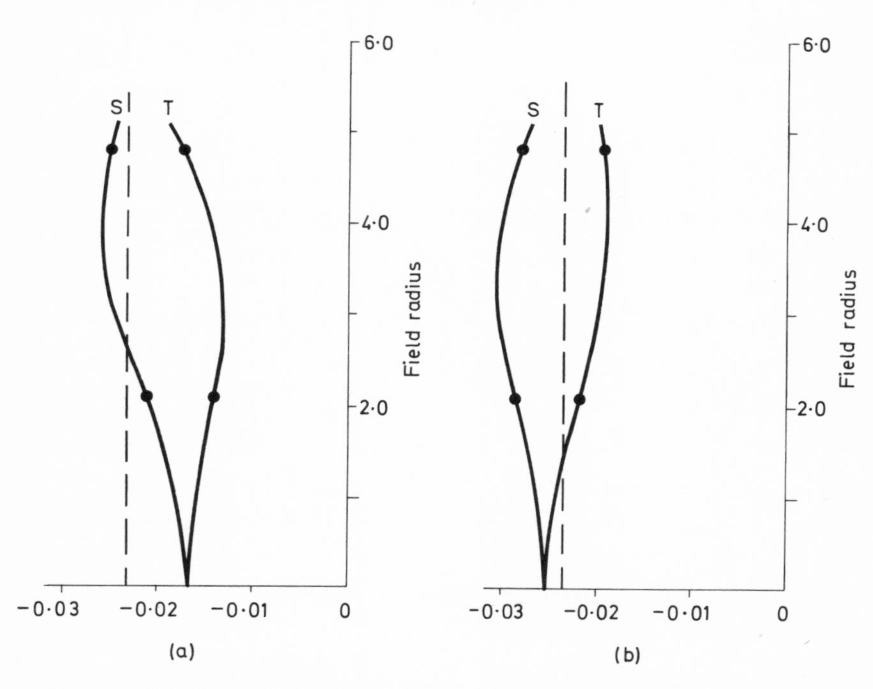

FIG. 33. (*a*) S and T field curvatures for best focus at 200 c/inch. (*b*) S and T field curvatures for best focus at 600 c/inch.

Some further information can be obtained from these curves by determining the positions of the peaks. These give the best focus for each field position and orientation, and may be plotted to give the S and T field curvatures for best focus at the two spatial frequencies (see Fig. 33). The position of best overall focus as determined previously is also shown by the broken line. These plots of field curvature give further indication that the OTFs at other positions in the field will not be very different from the two positions chosen. If the field curvature plots had shown large variations across the field then further calculations of the OTF at other field positions would have been necessary.

One other point which is of interest in Fig. 30(*a*) is that the tangential MTF at the larger field radius is significantly better than any of the others. This is due to the vignetting, which may be seen in the aberration curves of Fig. 28, cutting out much of the aberration. However, although this improves the contrast, as shown by the high MTF, the actual intensity of the image will be reduced by the vignetting.

We may now summarize the performance of the lens as follows:

1. The best overall focal plane under the conditions specified is approximately −0·023 inch from the focal plane.
2. At this best focal plane the MTF is better than 0·74 over most of the field at 200 c/inch, and better than 0·47 over most of the field at 600 c/inch.
3. At the best focal plane the lens is well balanced over most of the field since the MTFs for a given spatial frequency are equal to within a few per cent.

These figures do not of course give the quality to be expected from the final photograph, since no account has been taken of the transfer function of the photographic process. This is difficult to calculate but the majority of film manufacturers supply information which allows the contrast of the final image to be calculated from the contrast of the aerial image presented to the film and the intensity of illumination. Because of this it is of importance to know the illumination in the image produced by the lens over its working field.

This information is lost when the OTF is normalized at zero spatial frequency, and would usually be found by other means, as would the distortion characteristics which, as we saw in Chapter 4, are deliberately excluded from OTF calculations by choosing the ideal image point as the point at which the real rather than paraxial principal ray intersects the image plane.

APPENDIX 1

Discussion of Temporal Incoherence

The complex amplitude of a disturbance at a point P arising from a strictly monochromatic point source distant r from P is given by

$$K = \frac{a}{r} \cdot e^{\frac{-2\pi i \nu}{c}(r - ct)} \cdot e^{-i\phi}$$

where a is the amplitude
ν is the frequency of light
c is the velocity of propagation of electromagnetic waves
t is time
ϕ is the phase angle

Consider now two monochromatic point light sources s_1 and s_2 having amplitude a_1 and a_2, phases ϕ_1 and ϕ_2 and frequencies ν_1 and ν_2. The complex amplitude of the disturbance at a point P, distant r_1 and r_2 from s_1 and s_2 respectively, caused by the superposition of waves from these two sources, is given by

$$K = \frac{a_1}{r_1} \cdot e^{\frac{-2\pi i \nu_1}{c}(r_1 - ct)} \cdot e^{-i\phi_1} + \frac{a_2}{r_2} \cdot e^{\frac{-2\pi i \nu_2}{c}(r_2 - ct)} \cdot e^{-i\phi_2}$$

Thus the intensity Q at the point P is given by half the squared modulus of K (see for example Born and Wolf (1965), § 7.2). That is,

$$Q = \frac{1}{2}\left[\frac{a_1}{r_1}\cdot e^{\frac{-2\pi i \nu_1}{c}(r_1 - ct)}\cdot e^{-i\phi_1} + \frac{a_2}{r_2}\cdot e^{\frac{-2\pi i \nu_2}{c}(r_2 - ct)}\cdot e^{-i\phi_2}\right]$$

$$\times\left[\frac{a_1}{r_1}\cdot e^{\frac{2\pi i \nu_1}{c}(r_1 - ct)}\cdot e^{i\phi_1} + \frac{a_2}{r_2}\cdot e^{\frac{2\pi i \nu_2}{c}(r_2 - ct)}\cdot e^{i\phi_2}\right]$$

$$= \frac{a_1^2}{2r_1^2} + \frac{a_2^2}{2r_2^2}$$

$$+ \frac{a_1 a_2}{r_1 r_2}\cdot\cos\left\{\frac{2\pi}{c}(\nu_2 r_2 - \nu_1 r_1) - 2\pi(\nu_2 - \nu_1)t + (\phi_2 - \phi_1)\right\}$$

Since the frequency of oscillation of light waves is of the order of 10^{15} Hz and the fastest detector response times are of the order of 10^{-9} seconds, only the time-averaged value of Q is of interest, namely,

$$\langle Q\rangle = \frac{1}{T}\int_{-T/2}^{T/2} Q\,dt$$

$$= \frac{a_1^2}{2r_1^2} + \frac{a_2^2}{2r_2^2}$$

$$+ \frac{a_1 a_2}{T r_1 r_2}\int_{-T/2}^{T/2}\cos\left\{\frac{2\pi}{c}(\nu_2 r_2 - \nu_1 r_1) - 2\pi(\nu_1 - \nu_1)t\right.$$

$$\left. + (\phi_2 - \phi_1)\right\}dt$$

where T is the detector integration time, and $a_1^2/2r_1^2 = Q_1$ and $a_2^2/2r_2^2 = Q_2$, the intensities at P due to sources s_1 and s_2 alone. Thus, integrating, we obtain

$$\langle Q\rangle = Q_1 + Q_2$$

$$-\frac{2}{T}\sqrt{Q_1 Q_2}\left[\sin\left\{\frac{2\pi}{c}(\nu_2 r_2 - \nu_1 r_1) - 2\pi(\nu_2 - \nu_1)t + (\phi_2 - \phi_1)\right\}\cdot\frac{1}{2\pi(\nu_2 - \nu_1)}\right]_{-T/2}^{T/2}$$

Hence

$$\langle Q\rangle = Q_1 + Q_2 - \frac{2\sqrt{Q_1 Q_2}}{2\pi(\nu_2 - \nu_1)T}\left[\sin\left\{\frac{2\pi}{c}(\nu_2 r_2 - \nu_1 r_1) - 2\pi(\nu_2 - \nu_1)\frac{T}{2} + (\phi_2 - \phi_1)\right\}\right.$$

$$\left. - \sin\left\{\frac{2\pi}{c}(\nu_2 r_2 - \nu_1 r_1) + 2\pi(\nu_2 - \nu_1)\frac{T}{2} + (\phi_2 - \phi_1)\right\}\right]$$

$$= Q_1 + Q_2 + 2\sqrt{Q_1 Q_2}\cdot\cos\{2\pi(\nu_2 r_2 - \nu_1 r_1) + (\phi_2 - \phi_1)\}\cdot\frac{\sin\pi(\nu_2 - \nu_1)T}{\pi(\nu_2 - \nu_1)T} \quad (\text{A1.1})$$

Now, for temporal incoherence to exist,

$$\langle Q\rangle = Q_1 + Q_2$$

Thus the third term in equation (A1.1) must be zero or sufficiently close to zero to be negligible. In general, the cosine term will not be zero, so it is necessary to establish the conditions under which the sine term is zero. The sine term and the denominator form the sine function, sine$(\nu_2-\nu_1)T$, which is oscillatory, having its first zero at $(\nu_2-\nu_1)T = 1$. The next peak is at $(\nu_2-\nu_1)T = 1{\cdot}43$ and has (negative) height 0·217. Beyond $(\nu_2-\nu_1)T = 2{\cdot}68$ the sine function never again has a value greater than 0·10, which, with equal intensity Q produced by each beam and the cosine term in equation (A1.1) at its maximum value of unity, gives, for the time-averaged intensity,

$$\langle Q\rangle = Q + Q + 0{\cdot}1Q$$
$$= 2{\cdot}1Q \qquad \text{(i.e. a 5 per cent error)}$$

Assuming this to be an acceptable maximum error, one can determine the minimum separation in frequency of the sources for there to be no temporal coherence, namely,

$$(\nu_2-\nu_1)T \geqslant 2{\cdot}68 \qquad \text{(A1.2)}$$

If the integration time T of the detector is 10^{-9} seconds, which is conceivable with a suitable photomultiplier, then

$$(\nu_2-\nu_1) \geqslant \frac{2{\cdot}68}{10^{-9}}$$

or, putting $\nu_2-\nu_1 = \Delta\nu$,

$$\Delta\nu \geqslant 2{\cdot}68\times 10^9\ \text{Hz}$$

In terms of wavelength, this becomes

$$\Delta\lambda = |\lambda_2-\lambda_1| \geqslant \frac{c}{\nu_1\nu_2}\times 2{\cdot}68\times 10^9$$

Putting $\nu_1 \simeq \nu_2 \simeq 5\times 10^{15}$ Hz,

$$\Delta\lambda \geqslant 3{\cdot}2\times 10^{-8}\ \text{micron}$$

This separation is much smaller than the width of the narrowest available thermal gas discharge emission lines (e.g. $1{\cdot}36\times 10^{-6}$ micron for the 0·6438 micron cadmium line) although two frequency-stabilized lasers have been brought within this range and the beat frequency detected.

Note that equation (A1.2) also demonstrates that if one has a strictly monochromatic source, so that $\nu_1 = \nu_2$, it is necessary for the integration time T to be infinite for temporal incoherence. Thus to make a measurement of the OTF using strictly monochromatic light (if obtainable) would require an infinite time. So-called monochromatic OTF measurements are in fact made using quasi-monochromatic light (see Born and Wolf (1965), § 7.3.3), for which

$$\frac{\lambda_1 - \lambda_2}{\lambda_1 + \lambda_2} \ll 1$$

but the detection times are sufficiently long, as indicated above even for a narrow spectral line, to make the source effectively temporally incoherent.

Hopkins (1967) extended this discussion to include sources having a broad spectral range, and concluded that thermal sources can be considered as temporally incoherent.

APPENDIX 2

The Complex Amplitude Distribution in the Image of a Monochromatic Point Source

By consideration of the Fresnel-Kirchhoff diffraction formula and with certain simplifying assumptions it can be shown that the complex amplitude at a point P produced by a spherical monochromatic wavefront filling an aperture A in an otherwise opaque screen is given by

$$K_\lambda(\mathrm{P}) = C \iint_A \frac{\mathrm{e}^{ik\rho}}{\rho} (1 + \cos\theta)\, \mathrm{d}A \qquad \text{(A2.1)}$$

where k is $2\pi/\lambda$
θ is the angle between the normal to the wavefront at the surface element $\mathrm{d}A$ and the line joining $\mathrm{d}A$ to P
ρ is the distance from $\mathrm{d}A$ to P
C is a constant of proportionality

(See for example Born and Wolf (1965), §8.3.2.)

Now let the aperture in question be the exit pupil of an imaging system, and let the spherical wavefront be the idealized wavefront (reference sphere) producing an image at P′ (see Fig. 34). Provided that P is sufficiently close to P′ (i.e. within a few wavelengths) further approximations may be made to the integrand of equation (A2.1). First $\cos\theta$ may be equated to unity and the resulting factor of 2 taken outside the integral. Secondly, in the expression $\mathrm{e}^{ik\rho}/\rho$, $1/\rho$ may be considered a constant, whereas $k\rho$ in the exponent may not. This is reasonable since ρ will change by only a few wavelengths in, typically, at least tens of millimetres, which is a change of a few parts in several tens of thousands, whereas at the same time $k\rho$ will pass through several complete

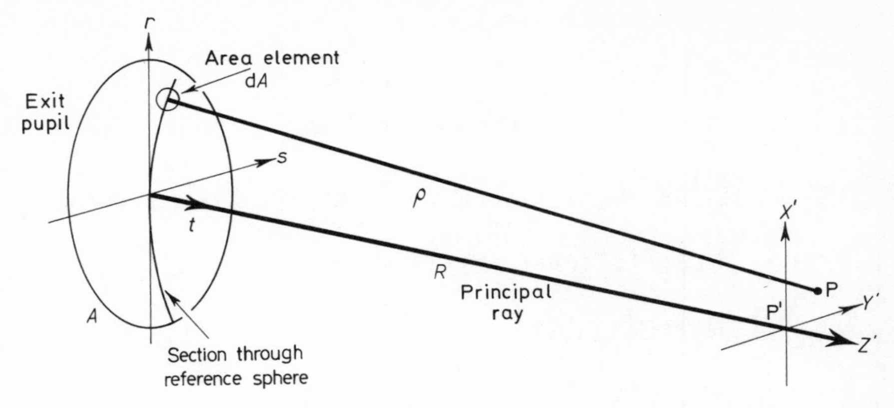

FIG. 34.

cycles. Thus the factor $1/\rho$ may be included in the constant of proportionality, and we have

$$K_\lambda(X', Y', Z') = C \iint_{\text{pupil}} e^{ik\rho}\, dA$$

If now the wavefront which fills the exit pupil is aberrated, then by definition there is a phase difference given by

$$e^{ikW_\lambda(r,s)}$$

between the reference sphere and the actual wavefront, where $W_\lambda(r, s)$ is the monochromatic wavefront aberration and r, s (and t) are Cartesian coordinates in the exit pupil as shown in Fig. 34. The complex amplitude at P is then given by

$$K_\lambda(X', Y', Z') = C \iint_{\text{pupil}} e^{ik\{\rho - W_\lambda(r,s)\}}\, dr ds \qquad \text{(A2.2)}$$

By Pythagoras,

$$\rho^2 = (r - X')^2 + (s - Y')^2 + (R - t + Z')^2$$

But $r^2 + s^2 + t^2 - 2Rt = 0$, from the equation of the reference sphere, and we can neglect $(X')^2$, $(Y')^2$ and $(Z')^2$ as being the second order of small quantities, giving

$$\rho^2 = (R + Z')^2 - 2X'r - 2Y's - 2Z't$$

i.e.

$$\rho = (R + Z')\left[1 - \frac{2(X'r + Y's + Z't)}{(R + Z')^2}\right]^{1/2}$$

Expanding this by the binomial theorem, to the accuracy required,

$$\rho = (R+Z')\left[1 - \frac{X'r+Y's+Z't}{(R+Z')^2}\right]$$

$$= R+Z' - \frac{Z't}{R+Z'} - \frac{X'r+Y's}{R+Z'} \qquad \text{(A2.3)}$$

The second and third terms of this equation may be combined to give

$$\frac{RZ'+Z'^2-Z't}{R+Z'} = \frac{-Z'(t-R)}{R} \qquad \text{(A2.4)}$$

since $(Z')^2$ may be neglected and Z' is small compared with R.

Now from the equation of the reference sphere, namely

$$r^2+s^2+(t-R)^2 = R^2$$

we have

$$\frac{t-R}{R} = \left[1 - \frac{(r^2+s^2)}{R^2}\right]^{1/2}$$

and expanding by the binomial theorem, to the accuracy required, gives

$$\frac{t-R}{R} = 1 - \frac{r^2+s^2}{2R^2}$$

Substituting this back into equation (A2.4), and from there substituting back into equation (A2.3),

$$\rho = R - Z' + \frac{Z'(r^2+s^2)}{2R^2} - \frac{X'r+Y's}{R+Z'}$$

and since Z' is small with respect to R this may be written

$$\rho = R + \frac{Z'(r^2+s^2)}{2R^2} - \frac{X'r+Y's}{R}$$

Equation (A2.2) then becomes

$$K_\lambda(X', Y', Z') =$$

$$C\iint_{\text{pupil}} \exp\left\{-ik\left(W_\lambda(r,s) - R - \frac{Z'(r^2+s^2)}{2R^2} + \frac{X'r}{R} + \frac{Y's}{R}\right)\right\} \mathrm{d}r\mathrm{d}s$$

Incorporating e^{ikR} in the constant of proportionality and including $-Z'(r^2+s^2)/2R^2$, which is the effect of defocusing by an amount Z', in the wavefront aberration term, gives

$$K_\lambda(X', Y') = C \iint_{\text{pupil}} \exp\left\{-\mathrm{i}k\left(W(r, s) + \frac{X'r}{R} + \frac{Y's}{R}\right)\right\} \mathrm{d}r\mathrm{d}s$$

which is the required expression.

Bibliography

Baker, L. R., 1955, *Proc. phys. Soc. B.,* **68**, 871.
Baker, L. R., 1965, *J. appl. Phys. Japan,* **4**, Supplement 1, 146.
Barakat, R., 1962, *J. opt. Soc. Am.,* **52**, 985.
Birch, K. G., 1961, *Proc. phys. Soc.,* **77**, 901.
Born, M., and Wolf, E., 1965, *Principles of Optics,* Pergamon Press, Oxford.
Bray, C. P., 1965, *J. opt. Soc. Am.,* **55**, 1136.
Bromilow, N. S., 1958, *Proc. phys. Soc.,* **71**, 231.
Campbell, F. W., and Gubisch, R. W., 1966, *J. Physiol.,* **186**, 558.
De, M., 1955, *Proc. phys. Soc. A,* **233**, 91.
Duffieux, P. M., 1946, *L'Intègrale de Fourier et ses Applications à l'Optique,* Besançon, printed privately.
Dumontet, P., 1955, *Optica Acta,* **2**, 53.
Elias, P., Grey, D. S., and Robinson, D. Z., 1952, *J. opt. Soc. Am.,* **42**, 127.
Goodbody, A. M., 1958, *Proc. phys. Soc.,* **72**, 141.
Goodbody, A. M., 1960, *Proc. phys. Soc.,* **75**, 677.
Herriott, D. R., 1958, *J. opt. Soc. Am.,* **48**, 968.
Herzberger, M., 1947, *J. opt. Soc. Am.,* **37**, 485.
Hopkins, H. H., 1950, *Wave Theory of Aberrations*, Oxford University Press.
Hopkins, H. H., 1951, *Proc. roy. Soc. (London) A,* **208**, 263.
Hopkins, H. H., 1953, *Proc. roy. Soc. (London) A,* **217**, 408.
Hopkins, H. H., 1955a, *Optica Acta,* **2**, 23.
Hopkins, H. H., 1955b, *Proc. roy. Soc. A,* **231**, 91.
Hopkins, H. H., 1957, *Proc. phys. Soc. B,* **70**, 1002.
Hopkins, H. H., 1967, *Advanced Optical Techniques,* North Holland, Amsterdam.
Hopkins, R. E., and Unvala, H. A., 1966, *Lens Design with Large Computers,* Proceedings of conference at the Institute of Optics, University of Rochester, N.Y.
Ingelstam, E., Djurle, E., and Sjögren, B., 1956, *J. opt. Soc. Am.,* **46**, 707.
Ingelstam, E., and Lindberg, P. J., 1951, *NBS Circular,* **526**, 183.
Jamieson, T. H., 1968, private communication.

KELSALL, D., 1959, *Proc. phys. Soc.,* **73**, 465.
KUBOTA, H., and OHZU, H., 1957, *J. opt. Soc. Am.,* **47**, 666.
KUWABARA, G., 1955, *J. opt. Soc. Am.,* **45**, 309.
LEVI, L., and AUSTING, R. H., 1968, *Appl. Opt.,* **7**, 967.
LEVI, L., 1969, *Appl. Opt.,* **8**, 607.
LINDBERG, P., 1954, *Optica Acta,* **1**, 80.
LINFOOT, E. H., and WOLF, E., 1952, *Mon. Not. roy. astr. Soc.,* **112,** 452.
MARATHAY, A. S., 1959, *Proc. phys. Soc.,* **74**, 721.
MARCHAND, E., and PHILLIPS, R., 1963, *Appl. Opt.,* **2**, 359.
MARTIN, L. C., 1966, *The Theory of the Microscope,* Blackie, London.
MIYAMOTO, K., 1961, *Progress in Optics,* Vol. 1, North-Holland, Amsterdam.
MONTGOMERY, A. J., 1964, *J. opt. Soc. Am.,* **54**, 191.
NAISH, J. M., and JONES, P. G., 1954, *Nature,* **173**, 1241.
OSE, T., and TAKASHIMA, M., 1965, *J. appl. Phys. Japan,* **4**, supplement 1, 154.
PALMER, J. M., 1971, *Lens Aberration Data,* Hilger, London.
POLSTER, H. D., 1955, *Rept. Perkin Elmer Co.,* No. 413.
SCHADE, O., 1948, *R.C.A. Review,* **9**.
THOMPSON, B. J., 1969, *Progress in Optics,* Vol. VII, North-Holland, Amsterdam.

INDEX